ROUX'S WAR

Also By

ADAM ROUSSELLE

THE MAHOGANY MAFIA

Conspiracy And Murder in The Jungle

COUNTING CROWNS

Catching Timber Thieves From Space

U|R|M|C

Revolutionizing The Energy Industry

SHORT CIRCUIT

Uncovering America's Largest Fraud

THE BASTARDS ARE BACK

Uncle Warren's 1951 Warning From Korea On Russian and Chinese

Infiltration and Today's Battle For America's Homeland Security

ROUX'S WAR

Profile Of An American Soldier

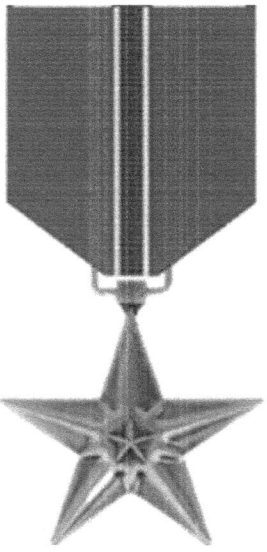

By: ADAM ROUSSELLE

"These are the times that try men's souls. The summer soldier and the sunshine patriot will, in this crisis, shrink from the service of their country; but he that stands by it now deserves the love and thanks of man and woman."

Thomas Paine

"My dream is of a place and a time where America will once again be seen as the last best hope of earth."

Abraham Lincoln

ADAM ROUSSELLE

CITIZEN SOLDER PRESS

2113 Middle Street

Sullivans Island, South Carolina 29482

Published simultaneously in Canada and The United Kingdom

Library Of Congress Cataloging-In-Publications Data

Rousselle, Adam 2024.

Roux's War: Profile Of An American Soldier

ISBN: 979-3-9920617-1-0

Series: Alacrity and Dispatch: Chronicles of a Citizen-Soldier's Selfless Service, Book One

Genre: Autobiography & Biography; Memoir; Action & Adventure; War & Military.

BISG Number (Book Industry Study Group)

BIO008000   Biography & Autobiography / Military

BIO026000   Biography & Autobiography / Memoirs

BIO038000   Biography & Autobiography / Survival

Library Of Congress Control Number (LCCN): 2024925571

United States Copyright Office Registration Number: TXu 2-454-374 August 31, 2024

Printed in the United State Of America

This book is printed on acid free paper

Book design by Adam Rousselle

THIS BOOK IS DEDICATED TO:

<u>My family</u>:
My indefatigable mother, Joyce
Catherine, Adam, Hayden, and Benjamin—
I hope this book helps you understand our lives better.

Gringo Joe

<u>My Brothers and Sister</u>:
Joey, Jim, Joe Jr.,

Dave, Rick, Billy, Kevin, Daniel, Josh, Brian

and Effervescent Eve.

<u>My Mentors</u>:

Les Thatcher; Jeff Bracker; Clint Vince and Steve Weber

<u>Veterans of Alpha Troop First Squadron 7th Cavalry</u>

<u>First Cavalry Division</u>

The men and women of America's Army—
those who do their duty every day,
bear the wounds of wars past,
and who died in service to our country.

In Memory of:

Staff Sergeant Jeffrey Bryant Mitchell

His marrow-deep Courage, only surpassed by his love for his Family

and his Soldiers

Why I Write: *The Promise of America*

I write because I believe in the promise of America—not just as a nation, but as an idea: that no matter how hard the road, no matter how impossible the odds, we can build a better world through courage, sacrifice, and an unrelenting will to serve others.

My experiences are proof of that promise. From the battlefields where I learned the true meaning of selfless service, to the jungles where I fought to reclaim justice for my family, to the courtroom and beyond—I've dedicated my life to standing up for those who can't stand for themselves.

Even if you don't know it, I've fought for you—and for a generous America that lifts up those who embody quiet courage, resilience, and hope; souls who fight their own battles every day—in small towns across this great nation—living the very promise our ancestors knew we could achieve.

Through the Alacrity and Dispatch series, I share stories of resilience, ingenuity, and hope. These books aren't just accounts of what I've done—they're testaments to the belief that one person, armed with truth and purpose, can gather teams of like-minds toward a common good and deliver extraordinary change. They remind us that the American Dream isn't just something you chase; it's something you fight for, no matter the cost.

I write to inspire you—to prove that no obstacle is insurmountable, no injustice unshakable, and no dream beyond reach. I write to remind you that American patriotism isn't about words—it's about action. It's about showing up, standing firm, and never giving up, even when the fight seems endless.

As you fight your own battles and chase your American dream, I offer a spirit—both American and human—that endures and overcomes. This is my way of delivering indefatigable hope that the best is yet to come for you and our Nation.

Table Of Contents

Chapter 1.   Get off the Mother Fucking Bus

I squinted to protect my eyes from the dust that flew around on that hot bus in the Georgia heat. I was on a "date with destiny" and headed for the Recruiting Command, which was the initial launch point for all new U.S. Army enlistees. I was excited, anxious, and kind of hungry. I was going to be an Infantryman. My recruiter told me that if you chose Infantry as your Army assignment you would become close friends with the guys in your Unit. He said, "You'll be with each other forever."

For a young displaced 18-year-old kid who never really had many friends, that sounded perfect. A chance to have some real friends outweighed the obvious hazards of potential combat. I hadn't grown up with dreams of joining the Army and going off to fight wars in foreign countries. I don't know what I dreamed back then, but I know that wasn't it. Hell, I wouldn't have even been there if my dad hadn't told me a few months earlier that "I just can't afford to feed you anymore."

He never kept food in the house. We ate at the job site and grabbed two sandwiches to take home. I couldn't understand why the lights were always out when we came home. He kept saying they were working on fixing them. So, we stayed out late, or rather he stayed out with some lady or other and I would find whatever trouble I could until I needed to go to sleep. The house had a waterbed for me. They are great unless they don't have electricity. Then it's like sleeping in a pool of cold water, draining your energy with every passing second. On the weekends I tried to hang out with folks who had lunch or dinner. I felt like a mooch, and I hated it. I didn't want anything for free, but I didn't have much choice.

It pained him to admit this to me and I can still see the look on his face. My Dad, Joe Rousselle, was a hustler. I'm not talking about drugs or gambling and trying to scam people. No. He simply knew how to make things happen and he always landed on his feet. He was successful at everything he tried and the house selling business was no different, until it was.

"I'm out of money Adam and don't know what to do, I can't help you anymore." He never minded being understated and matter-of-factly.

"Okay."

What else could I say? My dad had moved me from my home in Mechanicville, NY, leaving behind my mother, older brother Joey and younger sister   Eve. He had this notion that he was going to take me to Florida with him following my junior year in high school. He had the plan all mapped out. I would skip gym and lunch periods and graduate from high school a year early, and I would work with him in his business, we would build his business and I would learn a trade.

Never mind the fact that he had deserted our family years before when I was six. He was only a part-time parent, at best. But still he was my dad and he wanted me to come with him, so I had to go. I always loved him.

I ended up working in fast food restaurants (when we weren't trying his "next big thing") and taking too many classes with no real time to study. In the end, I graduated early with a 1.6 G.P.A., few good memories and a deep desire to do something with my life. As it turns out that something at this moment was going to be an Infantryman in the greatest Army on the planet and though I had no clue what that meant, I was excited about the whole thing.

So, there I was all 5'6" and 130 pounds of me, now standing in the middle of the Recruiting Command as people shouted at us.

"Hurry the fuck up!" I think I heard that 100 times as I stood in line, waiting to stand in the next line in order to get to the line to get a haircut.

It was a methodical and deliberate movement of bodies though each station here. Organized symphonic chaos. No one in charge cared about your feelings and your opinions about everything, including the heat, were unwelcome, and quickly shut down. We were being shown a glimpse into our immediate futures and for some this already seemed to be more than they bargained for.

But not me. I quickly realized that I actually liked it here. Structure, rules, and dependable treatment, whatever that treatment was, you could count on it.

When you first arrive at Recruiting Command, everyone has long hair. You could quickly assess who were the tough guys, who was rich, who was spoiled and who might be like you. People started to divide along those lines. But the Army knew this was going to happen. Everyone was forced into a line by their last name. This removed nearly all ability to find commonality with other bald people.

Bald heads removed nearly all individual distinction. I found that fascinating! "No more favored treatment for anyone," I recalled thinking. For a guy who always seemed to be on the outside looking in, this was heaven. I'd have a fair chance here. I didn't even care when someone told me, "Getting off that bus will be the worst experience of your life." I finally felt alive.

"Water."

I heard the voice before I realized that it was directed at me. It was another guy in my group whose last name ended with "R" or "S. That's how we were grouped together, all by last name.

Neither of us introduced ourselves, but he seemed like he knew more than everyone else.

"Drink water every single chance you get. Gorge yourself. If you miss the water, you will get weak, and they will pounce on you. Water will save you."

I told everyone his advice and reminded everyone to drink as much as they could. Most ignored me. "Why would I listen to you?" looks shot back at me when I spoke to them. I didn't mind taking advice or suggestions. I was like a sponge soaking everything up.

"Hurry up dumbasses!"

Non-Commission Officers, or NCOs, that's what they called themselves but we were told to call them Sergeants, constantly barked at us, making sure we were always on the go. But it didn't matter how fast we moved because all we were going to do was stand in line. And wait. That's the thing about when you first joined the Army. Everyone is always in a hurry to go nowhere. Somewhere along the line some big shot thought this was how the Army should be and it never really changed.

After our haircuts came the physicals, which can be summed up in one word, thorough. There was not an orifice on our bodies that wasn't checked. The Army scanned us for everything. Nervous tension was everywhere. It was like we were all racing towards the edge of a cliff and about to jump without knowing what was at the bottom. As I got fitted for my uniform, I felt a strong sense of pride receiving my new BDUs (Battle Dress Uniforms). They smelled like oil, perhaps formaldehyde, hazardous, yet somehow, sanitary. I rubbed my fingers across my newly issued dog tags. They had my name on them along with my blood type. I was told if I got killed in combat, they would place one between my teeth and tie one to my toe. This was how they would identify my body; I was a part of something now. And it felt oddly welcome.

We sat and waited for hours until everyone was done before we were all separated again. Bald, sweating, medically cleared and smelling like oil we had no clue what to expect next. It was 7 pm.

"Infantry! Left Line! Armor! Center! All other pussies that are not combat arms, Right Line." The Sergeant didn't blink. We hurried to our respective lines and waited for 6 hours. We were fed a box dinner and had bathroom privileges. Who has bathroom privileges? We waited until 1 am and then loaded onto the bus and our lives would change forever.

The drive to Fort Benning, Georgia seemed like it took forever, but it was only 7 hours. We were told to go to the bathroom before we left and didn't stop until we arrived. If you couldn't wait, you figured it out.

The bus stopped abruptly as if we had just crashed into a wall of solid Georgia heat. The door opened and a muscle-bound man with a weird looking, round, brown hat got on the bus. The vein in his neck bulged and you could almost see his pulse. Outside of the bus, 9 men similarly dressed stood. I thought they all looked like Tour Guides. I think the guy behind me actually said it out loud.

"Welcome to Fort Benning, Georgia's Harmony Church."

The Man's voice was a sharp contrast to his appearance. His tone was soft and calming. His smile seemed painted on, as if he had never seen a bad day in his life.

The Man looked down at the kid sitting in the front seat and flashed a bright smile of perfect straight, white teeth.

"Get off the bus," he softly said to the kid.

"Excuse me?" The kid looked confused by the command.

I think the vein in the Man's neck exploded. His eyes squinted and his face turned bright red. He grabbed the kid and threw him over the front seat rail and down the stairwell. I heard him as he hit the gravel parking lot with a dull thud. He then pulled the first three guys out of their seats and threw them over the same rail.

"Get off the mutherfucking bus!"

The backdoor opened and the men who had been standing outside watching us charged in.

"Get off the mutherfucking bus!"

They started throwing guys out of the backdoor. One hand on the collar the other in the small of the back. They were tossing guys like a sack of potatoes and sending them flying on the gravel.

I was in shock. I think we all were. These Drill Sergeants were sweeping through the bus and guys were flying in all directions.

"Are you fucking deaf? Get off the mutherfucking bus!"

I realized that whatever the Sergeant said, he meant. As they raced through the bus grabbing anyone not crawling over the next, I kicked out the side emergency window, and jumped out. Guys were screaming and crying for help. I had nothing to say about it. Nobody had anything to say about it.

These guys in the brown rounds were in charge and our lives and well-being were now in their hands. They were Drill Sergeants. But I wasn't afraid. I guess when you realize that you don't have any other options; there is no need to be scared. You literally have no choice but to manage whatever is confronting you.

These Drill Sergeants were Vietnam veterans. Three had visible wounds from their time in the War. Their heads were razor shaved and they were in a constant state of agitation. They yelled so much it seemed like their heads would fall off. There were no rules in 1984 about physical punishment. There were no women in the infantry and there were no Infantry Drill Sergeants without a purple heart and a medal for valor.

It was their duty to make sure we were well trained and equipped for the next combat assignment. They took their jobs seriously because they were tasked with teaching us how to save our own lives and those of our fellow soldiers. I respected them tremendously for that.

They pushed us into a formation and told us where to stand. The Senior Drill Sergeant then introduced himself to me in front of all the "children" (soldiers). He placed his hand on my shoulder.

"This guy's no idiot!"

He looked around as if checking to make sure everyone was paying attention when he knew full well, they couldn't be looking anywhere else.

"This little mutherfucker. What is your name Mr. Smarty Pants?"

"Rousselle Sergeant!"

I had learned enough to know that he needed to be referred to after everything I said.

"Roux what?"

"Rousselle Sergeant!"

"Fuck that! What lowlife Mama named you? How about I call you shell-shock, Rouxs? Yeah, that's right! Shut the fuck up"!

I had no intention of saying anything This guy was so close to me I could smell the eggs he had for breakfast. The spit flew from his mouth when he yelled, and it hit me in the eyes several times. But I wouldn't' dare squint.

"Why are you here?"

I wasn't sure how to answer this question, so I said nothing.

"I said why the fuck are you here?"

I thought for a moment.

"I'm here for the food Sergeant!"

I was being honest. My Dad had told me in no uncertain terms he couldn't give me any more money. I needed to eat.

The Sergeant looked at me. I think he cracked a smile, but it quickly disappeared.

"Listen Everyone! This shellshock douchebag, Frenchie piece of shit muthafucker is now in charge for the day. And he gets to eat last! Why does he get to eat last? Who knows why you lame ass pieces shit pussies? I'll tell you why, because you fuckheads need me to tell you why.

He eats last because only the leaders will know that their soldier's didn't get enough food to eat, when they eat last in line and there's no food. A good soldier doesn't complain. If there's no food, they just march the fuck on, they share with their fellow soldiers. When leaders don't eat, they can fix it so's everyone can eat next time. He's a fucking leader."

Food! I was starving. He could've called me anything he wanted to after that.

"You know what shellshock did to deserve this? He got the fuck off of the bus. This short mutherfucker mutherfucking jumped out of the window! That's the initiative we need around here! Fuck that window!"

I wanted to smile, but I figured if I did, he would take back his compliment. So, I did my best to look like a soldier. He sized me up. Then walked away to pounce on a kid who had made the mistake of asking one of the other Drill Sergeants a question.

I beamed inside. I had been singled out for taking the initiative. Yeah, I was going to eat last, but that didn't matter. That Drill Sergeant changed my life that day. He gave me confidence that I had never had before. I stood taller and straightened my shoulders a bit more. There was no turning back from here. In a matter of hours, I had already made a name for myself. I knew my mother would be proud and that made me feel good.

The foundation for my meteoric rise in the military had been laid.

My stomach growled.

I was unknowingly one step closer to my date with The Mahogany Mafia

Chapter 2. Smile and Wave

My dad left the house when I was six, I remember feeling bad for him when he left because I thought he would be so lonely. I had no idea how bad it could be after he left. After he left an emptiness would co me upon the house for the next decade until I joined the Army.

My older brother Joe, Joey, somebody I admired and looked up to always. I still do today. He was the handsome one that didn't wear glasses that had the cool hair and knew how to get along. The girls loved him. He was a jock, played in the Mickey Mantle Little League World Series. He was also the football player, he was my big brother and I always wanted him to be my friend; but his little brother was pretty geeky and that made it tough for him.

My younger sister Eve is indomitable and effervescent. Hopeful, beautiful and graceful, she was deeply impacted, as any young girl would be, by the perpetual absence of her father. Like myself, she was forever hopeful for his return and affection. However his exploits were a constant reminder that we were not even close to being first in his life. If I ever hated my father, it was for this reason alone. Of all things you might have broken, her heart was one precious masterpiece too many. I would never forgive him of this. I was too young to know how to help Eve and by the time I was, I was absent in the Army too. Failing my sister's needs has guided my view on how to be a better parent and husband.

I suffered from what I now know is attention deficit disorder and never understood the social and emotional cues that Joey understood and navigated masterfully. Beyond that challenge we lived very far from the school in an impoverished area, we had one car that was always on its last leg.

My mother, Joyce Moll Rousselle often resorted to siphoning gas out of the lawnmower in the winter to put in the car gas tank so she could go back to work and cut hair. I still remember mom spitting the gasoline out of her mouth one Sunday morning when she was filling the car tank with what little morsel of gas was left in the lawnmower. My mother worked tirelessly, and she shall forever be my hero.

My mom, Joyce Moll Rousselle, worked three jobs and we lived in a small area where her young sister Mary Ellen, a firebrand which cannot be extinguished, sought out the indignities of life and could challenge the status quo of a waterfall. Mom's sister Sylvia a redhead firebrand herself was a steadfast mother of three and a community rock. Always giving of herself and loving to me. Her husband Uncle Joe, was a great American. He worked on the B&M Railroad for 44 years and exemplified a small town father and husband who showed up, shut up and put up. He was at once imperfect, and dearly beloved, Uncle Joe was a rock of a man.

Her brother Warren was a quiet man who worked at the Orange and Rockland Utility. I would not know until August 1992 that Warren was a decorated 5th Cavalry machine gun platoon leader in the Korean war. His hand written, blood stained, letters home from Korea make the most hardened veterans weep when they read them. I will share his story in a non-fiction book honoring him and his fellow 5th Cav Troopers memory. Leland, Mom's brother became a Carpenter, I guess a consequence of his massive hands. Those hands would squeeze mine tight as a young boy and make me wonder, why. Stewart, himself a lifer on the railroad, was a gregarious storyteller and friend to all. He married Sandi and the two of them could throw parties. I never got to spend enough time with him, but appreciated the positive moments and kind words he always provided to me. Uncle Bill, he was handsome, a great golfer and had a selfish streak that taught me about things I could do better.

Sylvia and Joe were the most impactful relatives outside my mother. I don't know how they always smiled and found a way to have fun. I think they must've learned from my grandmother Beatrice. The stories that I would hear were that they were poor and when they would go camping, they would take some of the mattresses from the house and strapped them to the roof of the LTD station wagon, tie it down and drive to the park. My bride Catherine and I gave our son Hayden's middle name after them. Hayden Fitzpatrick.

Sylvia and Joe's children, my first cousins, were the closest external family members to me. Jimmy was closest to my age and was bright, charismatic and a dogged opponent at each and every game. We never stopped playing board games. I never stopped losing and Jimmy couldn't figure out why I never quit. To this day, this remains true. I would lose 50 times and ask for more. I guess I have always thought there are circumstances under which you could win and finding out how has been my life's journey. Joe Fitz Jr. Was a tireless worker and dedicated himself to being the best driver at UPS, before people even knew who they were. Joe exemplified work ethic and a steadfast father and husband. Joe once interrupted me when I introduced him as my first cousin: "Nope. Adam's my brother" Brenda! Brenda is like my sister Eve, indomitable. Brenda deserves respect and admiration for her perseverance and we share the same, no quit, no how, no where, marrow deep traits.

Grandma would share that when the nearby folks would smirk or laugh at the Moll caravan, grandma used to just look at the people who would stare at them and tell my aunts and uncles to smile and wave, SMILE AND WAVE EVERYBODY, smile and wave. Thanks Grandma Moll for your courage that carried itself to my core.

Indeed, smile and wave. My mother did. My mother was experienced, smart and she never quit when faced with adversity. She found her way through every trial and tribulation and insured that she had a smile on her face. I don't know to this day how she did it, affected by war and the things you read later in this book. I found my love for my mother growing more deeply as each passing day happens and it's with sadness at this late in my life I learned to appreciate so much about her. I imagine it must be a common feeling for those who pay attention too late in life. I feel glad that I know about it now and have told her so.

I Smiled and Waved when I arrived at every turn life handed or took from me.

I can recall so many acquaintances throughout my life always asked: Why are you so motivated? What is it with you?

I guess I thought everybody must act the way I was acting. I never thought I was anything other than less than common.

Chapter 3.   First Among Equals

"Are you out of your fucking mind Cue ball?" I knew where this was going. It happened every day. A Drill Sergeant would get pissed at one of us and it would go downhill from there.

"No Sergeant." The kid was visibly shaken, and this only incited the Drill Sergeant more.

"You talking to me?"

"No Sergeant."

"You looking at me? You want to kick my ass?"

"No Sergeant."

"You must be a mutherfucking soup sandwich mama boy thinking that you can ever even speak to me!"

I felt sorry for the kid. He looked miserable. That was the idea. Those Drill Sergeants were trying to tear us down and make us feel like the lowest of shit, only to build us back up and send us on our way as soldiers. Most of my unit looked miserable. They felt miserable. We were in South Georgia with 90-plus-degree temperatures daily and humidity that hung around 90%. Guys from New York and colder climates were suffering. But not me. I loved it here and I couldn't understand why anyone else didn't.

I didn't mind the Drill Sergeants yelling and spitting in our faces. I figured out early on that it was simply part of the process. I loved the structure of the Army. For a guy who had spent the last few months sleeping on a buddy's couch cleaning grease traps with no real idea of what to do after my dad's business failed, this was a chance at a fresh start. We were given everything we would ever need: food, shelter, clothing, shoes, training. I thrived by completely immersing myself in the culture.

If a Sergeant said they needed someone to volunteer or do something, I was the first to raise my hand. The latrine needs to be cleaned. I'd do it. The Connex needed to be cleaned. I'd do it. If it needed to be done, I did it and always asked "What's next?" when I was done.

I was 18 years old, naïve and dumb, but eager to succeed and willing to work hard. What I didn't realize at the time was I was establishing myself as a leader, a leader by example. And in the Army, that sort of thing doesn't go unnoticed.

I asked, "What's next?" so much so that one Sergeant started turning me down. He'd say, "Let someone else do it." Or "Not you Rouxs." The name had stuck with me, and I liked it. I made new friends and earned the respect of my NCOs.

That July, on my 19th birthday I unknowingly got a glimpse into my future. We were in a swamp just off the banks of the Chattahoochee River. It was hot as hell, at least 95 degrees. The humidity was like a hot wool blanket. We were belly down in the swamp, crawling, setting up an ambush during a training exercise. Water moccasins swam in and out of the dark murky water. It seemed like it should've been cool, but it wasn't. It was hot and steamy, almost boiling.

And it was only 11 am.

In 1984, the helmets we used in training were the same ones that were worn in Vietnam. When you're lying down, they don't fit very well, and the chin strap cuts into your neck. We had to lie still. No talking or swatting mosquitoes. No screaming if a snake swam across your arms. You stayed deathly still.

By 1 pm. the Georgia sun was baking the water and our backs. This was a test of wills, and I knew that I wasn't going to give up no matter how uncomfortable or hot I felt. Around 3 pm. I pissed through my pants into the swamp. I couldn't hold it any longer and I wasn't going to stand up. We had been instructed to stay down no matter what, because "if you moved, your friends would die."

I suddenly heard the loud barking of a Drill Sergeant ripping into a kid, who had decided to stand up and "stretch his legs." What the fuck was he thinking? "You don't stand up in the middle of an ambush exercise. I was angry that he didn't obey the rules. I was immature in most ways, but I recognized the severity of this decision and its deadly consequences had we been in combat.

"You stupid mutherfucker. You just got everyone killed!" The Drill Sergeant was direct. The soldier said nothing and I'm sure he got his ass kicked for what he did. I didn't feel sorry for him.

Years later I would find myself thinking about this day and those 9 Drill Sergeants. I didn't know this at the time, but they all exhibited signs of PTSD. They were hardened Vietnam Vets who undoubtedly had watched many of their friends die in combat in jungles in Southeast Asia. Their mental anguish manifested itself in highly aggressive and violent behavior. But still, they chose to put aside whatever their memories and torment may have been to help us learn how to save ourselves. You don't see that when you're going through it, but when your military duty is done, you have an opportunity to reconcile certain moments. I will always respect and admire them for what they did for me.

It was August 1st, and we were supposed to graduate Infantry School in mid-August. I got a visit from two stern-faced Sergeants. I was still in boot camp mode. I snapped to formation stance when I saw them. Two guys were standing about 100 yards away in civilian clothes, which was odd cause I hadn't seen anyone but soldiers since we arrived.

"Get your shit Rousselle. You're going to the Special Forces."

I almost fell over. How the hell could I be going to Special Forces? I was a Private E-1. Special Forces (S.F.) would never be an option for me, I was going to the 4th Infantry Division in Fort Carson Colorado as part of a Cohort Infantry unit. Hell, SF wasn't even going to be an option for anyone who wasn't a Buck-Sergeant E-5. I didn't ask them this and truthfully nor did I really care why. I was a soldier, and I did what soldiers do: I showed up, stood-up, shut up, and accepted my assignment.

"Yes Sergeant."

I was assigned to the 7th Special Forces Group at Fort Bragg, North Carolina. I was too young to fully grasp the weight of what had just happened: I left Infantry School before graduation and somehow still graduated. I was immediately granted a spot into airborne school. This itself was nearly impossible. Most guys had to wait 2 years to get in.

This was so out of the ordinary that the Commander of the 7th Special Forces Group's Service Company, Major "Mad" Mike Farrell, and his staff came from Ft, Bragg to Ft. Benning, Georgia to see who this kid was that they just gave a spot to in airborne school. They arrived and came to my room, and I wasn't there. I had taken the weekend off instead of heading to Ft. Bragg. He was incredulous. It was a dumb, immature, and naive move on my part. He left me a Green Beret on my bed with instructions to break it in so I wouldn't look like a cherry when I arrived at Ft. Bragg.

Though I had no idea what I was doing I arrived with the same enthusiasm. "What's next?" was my mantra and I maintained it.

"Who the fuck are you?" That was the question that everyone asked, and it wasn't because they were trying to get to know me. What they really meant was "How the fuck did you get here?

I didn't have the answer. I figured that's what happened when you worked hard and kept volunteering for things. The Army was always looking for volunteers. Hell, every soldier who wasn't drafted volunteered. But once you got in, there were still plenty of volunteer opportunities, so in my mind, I was there because I had earned it.

This was not the case.

My roommate was Sergeant E-5 Reynolds from Youngstown, Ohio. According to him, everything that came out of Youngstown was tough. He was a slob and kept his room in shambles. To top it off he dropped acid every day.

"You awe-da try it Man. It's good. It's good man."

He never told me what it was or what it would do. He showed me this piece of paper with these black dots on it. I didn't know what it was.

Reynolds was like a cross between a 70's Hippie and an 80's Headbanger. I couldn't figure out how he did his job. I was from Mechanicville, NY, I didn't know shit about acid. And I didn't want to. I didn't want it around me. I was scared of it. I was scared of him. I had to do something about it.

I went to see Staff Sergeant Bobby Sparks. Sparks was about 5'2" and African American. He was in charge of Supply. Supply had access to all kinds of shit. Supply had access to money. I don't care if it's a Submariner Rolex, like the ones that each of us got when we were selected to be Green Berets. The best timepiece on earth. The finest equipment on earth.

We had this big, locked cage. Everyone could go into this room. It looked like... well a cage and inside of it were god-awful amounts of capital-intensive assets that Bobby was in charge of, so if anyone needed anything, you get it from Bobby. And Bobby was always requisitioning stuff and if he got a signature from the right officer there was no limit to the shit that he could buy and the leverage that he had as a person.

Sparks was always nice and friendly to me. He liked me because he appreciated my willingness to volunteer for everything.

"Reynolds is dropping acid in our room every day." Sparks looked at me closely. I wasn't sure if he was about to tell me to "Get the fuck out of here" or what.

"Are you fucking serious?" This question was a vital part of every soldier's vocabulary.

Sparks sat back in his chair. He was thinking.

"Yes."

"Son-of-a-fucking-bitch!" Sparks was pissed.

He stood up. Sparks' wheels were turning. He told me to give him a day.

The next day I went to see Sparks again to get his decision.

"The Armorer is leaving. I'm going to make you the Armorer for the 7th Special Forces Group."

"I'm not an Armorer." I said that as if he didn't already know.

"You are now." Sparks spun his chair back around and turned his back to me.

I just stood there for a moment. In a matter of weeks, I had gone from being a Private in my second phase of Infantry training to Special Forces to the Armorer for the 7th Special Forces Group.

"Get out of here."

"Yes sir."

"And don't come around here with that formation shit anymore. You're not in boot camp."

The Armorer is in charge of maintaining all of the weapons and ammunition for the 7th Special Forces Group. Every MP5K, Every pistol; every gun. If it goes "bang" it only comes through the Armorer. He has the only set of keys to the weapons. The Commander can't even access it. The Armorer's keys and his barracks room had specially protected locks and keys and the Armorer's barrack's room could only be assigned to the Armorer. As a kid, I had always liked playing with rockets and understanding the mechanics of what it took to make them work. I was fascinated by the inner workings of the engines and other parts and not only how they worked, the way they worked, but why worked. Being the Armorer would be perfect for me.

Sparks had stood up for me because of my work ethic. Every time I would ask him, "What's next?" He gave me something to do, and he would smile and shake his head. He loved that I just got it done. He never told me to let someone else have a chance. Instead, he just gave me more stuff to do. Bobby saw me as "someone that goes along, gets along, and does what is necessary." He had gone to the captain and told him that he needed me to be his Armorer, because of my organizational skills and leadership.

By making me the Armorer, Sparks had guaranteed me an instant promotion. To be an Armorer for the Special Forces you had to at least be a Corporal. I had bypassed years of service and in an instant, I was promoted. I now had the biggest room in the hall and no roommates. I was going to Armorer school. The other Sergeants didn't like that. Many of them had held rank for nearly 3 years. Who was this kid that came in and became Armorer and now was getting special perks?

I was too young to see what was about to happen. This all took place relatively quickly. As I settled into my new room, I didn't know what to expect. I would soon have to go to Armorer school and learn all of the how's and why's of our weaponry. It was an exciting and thrilling time. For a 19-year-old kid, there wasn't anything bigger than this. I looked at the Rolex watch on my wrist. Months before I wasn't sure where I was going to get my next meal, but now I was on the cusp of becoming one of the youngest Green Berets in recent memory.

Everyone wasn't particularly happy about my meteoric rise though.

It was a Friday night and like most Friday nights we were gathering in the hallway to drink beers and talk. Ronnie Philpott handed me a Budweiser and slammed his beer. Ronnie was an interesting guy. He and his roommate were these two real-life cowboys from one of the Dakotas. They were bad asses. Strong and scrappy. The kind of guys you would want to take to war with you.

One day Ronnie and his roommate, whom he had known his entire life, got into a fight. No one knew for sure why, but they did. In the midst of the fight, Ronnie's roommate bit his dang ear off! Like literally bit it off. Ronnie bled and the two kept fighting until they eventually stopped and made up. Nobody messed with Ronnie Philpott or his roommate. They both were just too crazy.

I sipped the beer and Ronnie strode off. Girls were in and out of the hallway as they usually were on Friday nights. I was basking in the glory of my new promotion and feeling good about myself.

These three Sergeants were there with me. Everything seemed cool. Then all of a sudden one of them turned to me.

"The fuck you doing here" I had answered this question before. I didn't realize that I was moments away from an ambush. I laughed.

"You're a fucking asshole, you know that?" Before I could respond, they pounced on me. The three of them wrestled me into this sleeping bag. I tried to fight back, but it was useless.

"You're a pussy Rouxs!" One of them screamed at me. No one else on the hallway did anything. It wasn't because they didn't like me, it was because these guys were officers, and they couldn't. I remember seeing Michael Ways dip back into his room.

They quickly tied a Prussic Knot at the top and the bottom of the bag. Prussic Knots are climbing knots that tighten as weight is added. Climbers use them when they want to tie their sleeping bags on cliffs when they're doing multi-day climbs. They're reliable and strong.

They then tied some rope around the ends of the bag and secured it to the hinges on a door. I squirmed and kicked, but it wasn't going to do any good. I was stuck. The next thing I knew, they threw me out of the window! I felt like I was falling for seconds but in truth, I really didn't fall far. Maybe 7 feet. Then they left me to hang there.

I spent the night hanging high above the ground. I had hit my head on the wall when they threw me out, but I was conscious. No one came to rescue me or cut me loose. I spent the night hanging outside of that window, high above the ground.

The next morning, I missed Formation and Sparks asked where I was. He told the guys to go and untie me. I got out and went to Sparks' office; pissed off.

"They tied me up and threw me out of the window." Sparks just looked at me, then sat back in his chair like he often did.

"Get back to work." That was it. No one ever said anything about it again.

Why was everything happening so quickly? Was this how it worked for everyone? It couldn't be because everyone would've been in Special Forces.

I didn't have the answers. But I wouldn't need to look very far to find out the truth.

I did my best to try and fit in with the rest of the Unit. Not all the great soldiers at Service Co. 7th group were going to be Green Berets. Frankly, 3 in 10 of them attempted to get into the Q-Course. MSG Hebert told me he had only met two soldiers like me in his 27-year career; one died early, and the other was a division commander. He said I had an equal opportunity to be either. He told me I was special and to act like it. I applied and got accepted into the Special Forces Assessment and Selection Course (SFAS). SFAS is a short 3-week course that weeded out the sick, lame, and lazy. Pass this and it gets even harder. I wasn't big or muscular and I can't say that I started at the top of my class. But I wasn't the worst either. I held my own and quickly established myself as someone who had excellent navigation skills.

Those skills were about to take me to the jungle, and I didn't even know it.

Chapter 4.   Gringo Joe

My Dad left our family when I was 6 years old. My brother, Joe, and I were out-side riding on the dirt bike he had bought us. It was perfect for us; not too big, not too fast. At that time our family enjoyed a comfortable life. My Dad worked for a big drywall company and had bought a large house for us on this expansive parcel of land. We didn't really have neighbors so most of our social activities took place on the family's land. We were happy until that day.

"You boys take care of ourselves." Joe and I looked at our dad and nodded. We assumed that he was telling us to be safe on our dirt bike. He climbed into his car and drove away. We hopped on our dirt bike and continued riding, not knowing that our lives would soon change forever.

When Dad left, we soon went into abject poverty. My Mother had to work three jobs just to try and take care of us. We wore our shoes until they fell apart and our clothes were worn and often too small. Still, my mother always carried herself with a certain dignity that truly affected me.

"Be sure to always be kind to others," she would say, smiling through the pain that she felt in her own life. She exhibited this strength and this spirit that I be-lieve strengthened me as well. My Grandmother had this saying that she would always tell us when we were children, and my mother exhibited this mentality with her very being.

"No matter what, smile and wave." That was my grandmother's charge. She told us to never let people see what you might be going through. The saying was meant not to hide what we went through, but rather not be embarrassed by it. Kind of a "Let them stare. They have problems as well and we shouldn't shrink from our own reality. We are good people and should always be proud."

Food was scarce and we ate mostly soups. It's amazing how much we take food for granted until we realize that we don't have any. The real-life hunger pains that I felt were a gnawing agitation that I will never forget. I had no real friends at the time, so there was no real escape or a place to go to get a meal. What we had was all that we got.

Dad would call us once a year. Occasionally he would come and visit. Most of the time he would bring his girlfriends with him. I never knew how my mother felt about this and she never said anything to us kids. Dad was a handsome gregarious kind of personality. He loved women, drinking and eating. People loved to be around him, and he enjoyed their company. He just had this way with people.

We didn't know where Dad was living or what he was doing. We didn't know why he left, and I never found out. I never asked questions and he never volunteered information. We just left it at that.

Eventually, my mother had enough of our lack of food. She got my dad on the phone and I saw her angry for the first time.

"We don't have any food, Joe!" I was standing close by so I could hear my dad on the other line.

"Come on. It's not that bad. Stop exaggerating."

My Mother handed me the phone and I untangled the long chord and put the phone to my ear.

"Tell him" She said.

"We don't have any food. I haven't eaten in a couple of days."

My Dad listened. I handed the phone back to my mother and she hung up.

The next day this moving truck arrives, and these four guys get out and begin unloading this truck full of food. They didn't look happy about being there, but they did their jobs. We got everything we needed or wanted. I'm not sure how my dad did it but he fixed the problem. Food was never an issue again for the next 3 months. I don't know who those guys were or where they came from. At the time I had no idea where Dad might be.

In true Dad fashion when I turned 16, he showed up at the house with a brand-new Trans Am for me. What? Never mind the fact that he had been absent for most of my life. He just showed up like nothing was strange or different about that. That Pontiac became my escape and my ticket to finally getting to know some people.

My brother Joey had always been the popular one. He was the star football player and all-around great athlete. Good looking and outgoing, everyone loved him. He was like my dad in that way. I idolized him and that Trans Am made me feel cool like Joey.

Perhaps out of guilt or some level of remorse, Dad decided that I should come and live with him since I didn't seem to be doing much in New York anyway.

After the collapse of the Housing Market in 1984, my dad lost everything, which is subsequently how I ended up in the Army. But he wasn't done. He was smart, cunning, and entrepreneurial in spirit. He wasn't about to let that stop him from his next endeavor.

He ended up being quite successful at it and made enough money to sustain himself and live the life he wanted to live. "Gringo Joe," as the Hondurans called him had moved to a foreign country and carved out a niche for himself. The people loved him, and Honduras had everything that Dad ever wanted: women, alcohol, parties, food, and more women.

Soon Gringo Joe crossed paths with a very wealthy and influential woman named Argentina de Batras. Her family was Honduran aristocracy and she fell in love with and subsequently married my dad. And for all of the heartache it takes to say, it was clear he genuinely loved her and was faithful to her as well. I cannot reconcile my sadness that his love wasn't destined for my mother, she certainly deserved the love she needed.

Argentina was highly religious and a practicing Catholic. She insisted that my father prove that he was divorced from his previous wife(s), so she did not have to live in shame. While My father was not technically divorced, he insisted that he was. He was certainly devoted and true to her otherwise. They lived in that cycle of false reality, although in fairness to both of them, I am not certain she was aware.

Her family objected to Argentina marrying Gringo Joe, but she wasn't leaving him, and he wasn't leaving her. Their marriage thrust my father into the center of high society circles and culture. His larger-than-life persona and infectious smile were welcoming to all and intriguing.

"Who was this American gallivanting around with Argentina de Batras? Was he an American Spy or Special Agent? Was he there to do harm.? Dad's parties and mixers brought him face-to-face with some very famous and infamous people in Honduras, and he didn't even notice.

The U.S. Intelligence community noticed, however.

They had their eyes on this portly, carousing American and wanted to find out who he was and who he might be working for as well.

The Civil War in Nicaragua had finally ended, and it was known that the United States led by then-president, Ronald Reagan were supporters of the Contras who had opposed the Soviet-backed Sandinistas. This was a wider vision of Reagan's desire to keep Latin America free and open for U.S. defense and future trade opportunities.

Honduras and Nicaragua are neighbors and when the war ended many of the contras had moved into Honduras. The U.S.'s interest was to ensure that the Hondurans were trained and prepared for any potential Sandinista advancements into the country. Unbeknownst to Dad, Argentina de Batras' family and friends were also supporters of the Contras and thus there was a particularly high level of interest in figuring out who was my Dad and what was Argentina and her family and friends really plotting to do.

Dad was in the middle of an international surveillance operation and he was completely clueless.

One day as he sat at a swanky bar in Honduras another American cozied up next to him and the two began to talk. Dad loved conversations and never met a stranger he didn't like. He was a world-class schmoozer, so he took full advantage of the chance to talk to another American and perhaps even embellish some details of his life. The Man was vacationing in Honduras and he had plenty of time to listen to Dad's stories.

At some point in the conversation, Dad revealed that he had a son, me, who was a member of the U.S. Army. This news piqued the stranger's interest. Dad told him that I was in Ft. Benning, Georgia at the time. The two shared a few more laughs and drinks and Dad eventually left. That was August 1984 while I was in Infantry School in Harmony Church.

That was not a chance meeting.

The Man Dad was talking to was none other than Les Thatcher, the G2 of Army Intelligence. He and his men had been observing Dad for some time and he decided to get a personal feel for this Joe Rousselle guy. Dad unknowingly had walked right into a meticulously devised intel operation. Fortunately, he seemed to be on the right side of this battle.

Still, Les Thatcher apparently wasn't too sure.

Dad may have been thought to be a patriot, but he was gallivanting with some infamous people in Honduras. Les decided that he needed to keep a closer look on Joe so he picked up the phone and made a call.

"Who the fuck and where the fuck is Private Adam Rousselle?"

I imagine the person on the other end couldn't answer fast enough.

"I want him down here in Honduras."

And that was it. My military career was immediately thrust into a fast-tracked vortex, because Les Thatcher, an American intelligence Legend, had decided he needed me to spy on my Dad and his friends in Honduras.

I didn't know anything about this at the time. Les Thatcher would eventually become a mentor and advocate for me throughout my military and civilian life. He liked me because I was genuine, talked frankly and he knew I was a kid that needed help. Les Thatcher felt that I could be molded not just because of the potential information that I could provide, but I think he saw some bit of himself in me. I was a go-getter and he respected that.

As I continued going through the SFAS with my Green Beret wanna-be's, I began to get a better sense of the situation that I was in. I was training to become a Green Beret. A member of an elite Special Forces Group. There are only two ways you can become a Green Beret: You either fight with them in combat or you go through the Q Course. Les Thatcher had demanded that I go through the latter, despite my inexperience.

But I was changing. I was still only 19, but I started to mature a bit. After I snitched on the guys who threw me out of the window, Sgt. Sparks had made it adamantly clear that snitches in the Army weren't looked upon too highly. This revelation had set me straight, in a sense.

I needed to take this assignment seriously. I wasn't like anyone else. I was special. Someone, somewhere thought that highly of me and I needed to do my part. I worked hard and overcame my physical limitations to become 3rd in my class. I had my sights set on continuing to climb in the ranks..

One day, as I was going through our cage, Sparks approached me and handed me a small notepad and a pen.

"Put this in your pocket. Don't ever lose it. Carry a pen and a piece of paper with you at all times."

I tucked the small notebook in my pocket and thought nothing of it, but that little nugget of advice would change my life forever.

I had not been at Bragg long and just started the SFAS course when Command Sergeant Major (CSM) Peg-Leg Pete Parker approached me one day. He was an impressively short man, Peruvian and a stone-cold killer. He was the Command Sergent Major of the 7th Special Forces Group, and he only had 1 (one) leg. He was one bad ass dude with a gentle smile that didn't fit his gravitas.

My sergeant major always had great advice and listened when people spoke to him. He knew this news was about to change my entire career and he was simply the messenger.

"You're leaving." I was uncertain of what that meant.

CSM Parker seemed pained by the next few words. "You're going to be operational." His disgust couldn't be hidden. Vamos Ahora.

"Si Sargento Mayor."

How the hell was I going to be operational? I hadn't even finished SFAS training.

Les Thatcher and the CIA had decided that I was to be an operational asset and I needed to be in Honduras as soon as possible. I didn't know it then, but I would never finish training or earn a Green Beret.

CSM Parker told me I was assigned to Operational Detachment Alpha (ODA) 735. This was a scuba team in the 7th group, currently assigned to mine the harbors off the coast of Nicaragua.

I had never scuba-dived in my life and I wasn't a swimmer.

Chapter 5.   Honduras Assignment

I didn't know what to expect when we departed "Green Ramp" Pope Air Force Base in Cumberland County, North Carolina. Everything had happened so quickly, that I didn't have time to be nervous or even scared. As I tried to sleep on the C-141 Starlifter, I imagined how proud my mother must've been, knowing that I was on some sort of mission in the Army.

Mission? It didn't seem like much of one to me. I knew that our base of operations would be in La Ceiba, Honduras at Goloson Air Force Base and we would be deploying to LeVente and Mocoron where we were training the Contras. I didn't know much about La Ceiba and only had cursory knowledge about the Contras. Oh, and I knew that I needed to keep a close eye on Argentina, my dad, and their circle of friends.

La Ceiba was this beautiful, tropical port city just off of the Gulf of Honduras. It was a tourist town that attracted people from all over the world; especially Americans who had decided that this Central American/Caribbean paradise was the perfect place for them. It was the most beautiful place I had ever seen, and despite the excitement of my new assignment, I felt at ease and at peace there.

The people in La Ceiba were generally friendly and, in my role of working with the intelligence officers, I began building relationships with the locals that would serve me far after my years in the Army. I learned early on that the best way to endear yourself to a wealthy and powerful Honduran was to be nice to his wife and children.

This could easily be done by providing simple gifts that I could easily get through the Army. Practically anything that I asked for, my Intelligence managers would make it happen. Teddy Bear. No problem. Sunglasses. No problem. Perfume. Done. I remember one of the guys' wives coming to me upset. Her daughter was having a baby shower in a week and she needed some of those colorful onesies to give to her. The problem was there weren't any onesies anywhere in La Ceiba.

"Adam can you help me?" Her plea was heartfelt and troubling to me. I genuinely wanted to help her and do this for her family.

I went to my Officers.

"I need onesies for a baby shower. It will go a long way for us." I was telling the truth because it would.

A couple of calls were made and the next thing you know is she's got onesies in damn near every color you could think: light green, pink, yellow, sky blue; you name it.

The look of gratitude on her face made me feel good, but when her husband found out it was something different altogether.

"You got those onesies for my wife for my daughter?"

"Yeah it was no big deal."

He smiled big and shook my hand.

"Anything you ever need, you let me know."

Mission accomplished. I had gained an ally for life.

Nothing seemed out of the ordinary with Argentina. She was like every other rich person in the region. She enjoyed lavish parties and events, but there was nothing amiss. Still, I reported to my superiors, unsure of whether or not my work was useful.

Argentina was always kind and generous to me. I admired her strength and the way that she carried herself when she walked. I often imagined that if my

Mom had been born into wealth and exposed to so much opulence, she too would've been the host of town. Whatever it may have been.

My Dad couldn't have been happier to have me with him in Honduras. He bragged to his friends and anyone who would listen about me being a Corporal in the Army and the work that I was doing.

"He's in charge," he would often tell a stranger. One look-over at my youthful facial features and the motorcycle that I rode into town from the base, would quickly disprove my Dad's assertion. He knew that and the strangers probably knew that he knew that, but he didn't care. He was Gringo Joe.

Truthfully, my Dad was unknowingly on to something. I was a Corporal in the Army at the time and being a Corporal meant something. It was a special thing. It had a cache' that military people appreciated. You only make Cpl. as a lateral promotion from the rank of Specialist. It meant that the leaders around you, identified you as something different, and someone worthy of being a noncommissioned officer a year ahead of your contemporaries. Everywhere you went as a Corporal people would look at you differently. They would say hello to you in a way they would not say hello to other soldiers. Even the real hard-core Green Berets looked at me differently. They didn't know who the hell I was, nor where I had come from, but they knew I was a Corporal. Most of them barely spoke to me, but they would always pat me on the shoulder. Nod to me. They were 35-year-old men and they were killers, they even frightened me, but I always knew that I had the respect of everyone and I felt part of the team.

I didn't believe that I was spying on my Dad and on more than one occasion I considered telling him, that the U.S. Government was tracking his girlfriend just to be safe, but I didn't. He probably would've thought it was funny or maybe even cool, and I know he would've told everyone he knew that he was "under U.S. surveillance.

Seeing my Dad's happiness in Honduras was bittersweet for me. I still harbored angry feelings towards him leaving our family and even telling me to leave. Every time that I spoke to my Mother, I couldn't help but think of her sadness and loneliness. Here he was thousands of miles away in what some may call a Central American paradise, living it up and galivanting around town like an aristocrat, while she was at home probably working.

I spent most of my free time on the beach thinking about my Dad, Mom and Brother and Sister. Up until then, I hadn't thought that deeply about anything. Everything was about the moment and survival. But La Ceiba can do that to you. It can make you dig deeper into the depths of your Soul.

"It's so great to have you here Adam." My Dad looked earnestly at me. The day had gone like so many others and now as I prepared to head back to base, he seemed livelier than ever.

"It feels good to hear," this was the truth. I loved Honduras.

"How long do you think you'll be in?"

Hell, I had just joined the Army and my Dad was already asking me when I was leaving. I knew that his wheels were turning, because they always were.

"I don't know. For as long as I want to."

"Good answer." He stepped back from me and watched a car zip by.

"Don't stay too long though. There's a lot of opportunity in Honduras."

I thought about this for a moment. My Dad was always talking about "a lot of opportunity" somewhere. There was "tons of opportunity" in Florida, but it evaded us. There was "opportunity" in Mississippi, Tennessee, New York and every other state he had visited. None of those opportunities worked out particularly well in his favor, but his fire wasn't diminished.

He was right though. There were tons of opportunities in Honduras, you could just feel it.

"This is where WE need to be." He spread his arms wide for emphasis.

"There's a lot of business here Adam. Waiting for us. All we have to do is seize the moment at the right time"

My Dad looked at me. His eyes told me this wasn't another drummed-up scheme or loose idea. He had truly put some thought into it.

And then, he smiled and turned abruptly towards the house.

"We'll figure it out."

"Okay."

I hopped on my motorcycle and fired it up.

Two days later after 6 months, 15 days, 9 hours, and 25 minutes, my surveillance work in La Ceiba ended. The CIA and Army Intelligence guys had determined that Argentina nor my Dad posed any threat to the United States. I think in my heart I was relieved. I never thought my Dad or Argentina had anything nefarious going on, but I was glad that this whole thing could officially be closed.

Now, I would go back to repairing Goloson Air Force Base and running missions to Mocoron and LeVente.

Little did I know that my "opportunity" would be right in front of my face.

Chapter 6. Why Are They Throwing Rocks?

Getting to Mocoron and LeVente wasn't easy. You had to either have a helicopter or you needed great weather. We didn't use helicopters for most of these missions, so we had to make that drive. It was 6 hours through the jungle and mountains on a good day if the river didn't wash out the roads or the monsoon rain didn't turn the whole road into a mud pit.

On these trips, I would ride on the back of this 2.5-ton truck, a "Deuce-And-A-Half," as I called it, soaking up the world around me. We traveled with loaded weapons and military equipment to those forward bases like we were riding through a war zone. This was my first mission. This was my norm. I remember thinking that all soldiers carried loaded weapons whenever they traveled. I just didn't know any better.

The sun would beat down on the back of my neck (when it wasn't raining) like I was standing on top of it. But everyone was hot and I didn't complain.

There was growing sentiment that we needed to keep a vigilant eye on Sandinista activity near the Honduran border. There was always this fear that the Sandinistas might try and seize control of the Honduran government, so maintaining the current government was always an important part of President Reagan's strategy in the Region.

This prompted us to spend more time closer to the jungle than at Palmerola AFB, the U.S. Military base in Tegucigalpa, Honduras. Tegucigalpa was nothing like the coastal city of La Ceiba and about 4 hours away by truck on winding mountain roads. Each morning we would wake up and drive along these twisting dirt and gravel roads, overlooking lush valleys with fast-moving rivers.

I paid attention to everything that I saw, though I didn't realize that I was sub-consciously taking stock of a future that I had never imagined.

The giant trees hung like canopies over sections of the road; sometimes block-ing the hot rising sun for miles at a time. Thank God. It gave me a reprieve from direct sunlight I had grown accustomed to.

So there I was, on every mission, sitting on the top of the Deuce-and-a-half with a crew-served weapon, (that's a weapon so big it takes at least two people to use it). Riding through the jungle and mountains in Honduras, every trip seemed different to me. I always paid attention to small details so I would see new birds and vegetation every time we went on those road trips. I often thought to myself how lucky I was to be there at this time. To be able to gain this experience.

One day though, something completely different happened that altered my entire military experience. It must've been around 9 a.m. and we had been driv-ing on this road for about two hours. Every 5 to 7 miles or so we would come across small villages and the children would come running from their huts with big smiles on their faces, awestruck by our truck. They'd wave at us and I remember looking at them and being overwhelmed by just how happy they were. They didn't have much, but it didn't matter. They were genuinely happy in the middle of the jungle.

This continued as we rode through other towns and villages. It wouldn't just be children, adults would come out too; happily waving and smiling at us. You would've thought we had just saved them from some unknown fate. I always smiled and waved back.

We never stopped to say hello. We either weren't allowed to or frankly, we knew better. We were only miles from Nicaragua, and we were training the

Contras. The Contras were having an active war against the Sandinistas. If you were there, you knew this.

After a while, I would see fewer adults accompanying the children. The kids would still wave and once in a while the adults would.

A few miles later the children were not on the street, but the adults were, and they just stood there. I just figured maybe they'd seen the military trucks before and it was no big deal.

After another half an hour passed, it was more of the same, but then it changed. Instead of just standing there, they would turn their back to the truck and cross their arms.

These people didn't like us!

But why? We hadn't done anything to them. I made mental notes.

A few miles later instead of turning their backs and crossing their arms, they just had their hands outstretched with her middle finger extended and pointing to the ground. In America, we turn our hand up with our middle finger extended and give people "The Bird." In Honduras, they point to the ground and wag their hands in a side-to-side way. They were giving us the Honduran bird!

A few miles later, shit got real. A man picked up a rock and threw it at us and hit our truck as we drove by!

"Keep your eyes out!" I pointed to the driver and he nodded in agreement.

Sightseeing aside, my job on the top of the truck was to provide security. I took my weapon out of safe for the first time. I actually felt kind of silly and I didn't know what to do. I was a young kid on the top of that truck. But people throwing rocks at you in the wilderness, will change your attitude in a hurry.

In a few hours, we had gone from waving and smiling children to pissed-off people throwing rocks.

We continued driving toward our forward operating base and the behavior happened in reverse. At first, more people would throw rocks at us. As we got closer to the base, we'd then get the Honduran "Bird." Then people turned their backs to us and folded their arms. Then some just stood there. As we got closer and closer to the base, we would see smiling faces and waves from every-one. Late in the afternoon, we arrived at our destination and I didn't know what to make of it all.

"Why the fuck were they throwing rocks!"

When we went back to Golson Air Force Base the same shit happened again. This happened the next day and then the next and the next. I had a hunch about the whole thing and decided to test it. I found a small road where we could take a different route and approach the center of the area where they were throwing rocks from a different direction.

We began our new route with children smiling and soon found that the chil-dren were being called inside by the parents. Then the parents would cross their arms and turn their backs until eventually, someone would throw a rock.

What the hell are these idiots thinking and why are they throwing rocks at US?

We were there to help the people better their lives. We were there to support them and help with their livestock and farms. Clearly, they thought otherwise, even though our actions spoke differently.

This shit went on one too many times and I decided to start keeping track of where each change in the people's behavior occurred. I wrote down my notes in the green notebook that I carried with me everywhere. I had always been an excellent navigator and I genuinely enjoyed that sort of thing. I drew a crude map in my notebook and began plotting where the rocks were thrown and changes in the villagers' behavior.

Maps are the soldier's language. Everything we did back then was driven by data, information and geography. I mastered navigation early on and listened to the advice from Staff Sergeant Bobby Sparks to "Never be caught without paper and a pencil. Someone could die." I put our knowledge to work.

When I was done drawing my map and marking our experiences on it, I took my notepad and tore off the four little pages from its mini binder, and I gave it to command Sgt. Maj. Peg-Leg Pete Parker.

The next day I was called into a makeshift barracks in an abandoned building at the Fourth Infantry Battalion outside of Le Ceiba. When I walked up to the building six young men were wearing what looked like white diapers, with their hands bound behind their heads with leather ties and a steel railroad track through their arms behind their heads squatting together trying not to fall over and break their arms.

They were the Castigados. The outcasts of the Honduran Army.

The Castigados are soldiers who for some reason, did not do what they were told and the punishment was well, corporal: violent dehumanizing, and public.

The Honduran Army had far different rules than the U.S. And it started with their recruiting command. It was this wonderful mobile, school bus, that had nice camouflage green paint on both sides and it said Army in Spanish on its side.

The recruiting team would drive into the city center in Le Ceiba in Parco Central. The recruiting bus would then stop in front of the movie theater. This was the same movie theater, across from the catholic church, that had a free matinée that day and all the young boys and all of the young girls would go to the matinée together. The soldiers would get out of the bus walk into the theater and beat, all of the young men with billy clubs out of the theater and into the bus.

Welcome to the Honduran army.

The Castigados were a sign to everyone that the Honduran military were serious. There was no law but their law. Being a United States Army soldier gave us rights and privileges and protections that created animosity between their enlisted soldiers and ours. We were never worried we'd be punished like that, never could they disrespect us.

I walked into the building. The Sgt. Major asked me to sit down with these two guys, who were wearing civilian clothes. They were working with my notes.

They had new maps on easels and stacks of yellow Manila file folders filled with satellite photography and more photos up on the wall.

One of the men looked at me.

"Adam…" I was shocked to hear my own name. I hadn't spent much time around civilians since my intelligence gathering days with my dad. Everyone called me "Corporal" or "Soldier." I knew he was serious.

"Who are you?" I asked.

"You're imagination," the older of the two gentlemen replied.

I knew what that meant. This guy was a big fucking deal, and he wasn't here for pleasantries.

"Great fucking work Adam. Half of the mutherfuckers around here might've thrown rocks back at the fuckers."

The older Man held up my notes.

"Why did you write this?"

I looked at command Sgt. Maj. He nodded approval.

"Someone in the middle doesn't like us," I told him.

I pointed at the map where the roads intersected.

The two civilian-clothed men didn't make a sound.

"That seems pretty obvious," I added.

"To whom?" The old Man asked.

I wasn't sure how to answer that, but he saved me.

"Actually it's not obvious at all. What you have done is a sharp and rare assessment. You have identified the location of a bunch of fucking assholes who are trying to fuck with the United States of fucking America. And we really don't like to be fucked with."

The old Man studied me. Looking through me, as if deciding what he wanted to do with me.

"The name's Ness." He looked me square in the eyes.

I had been lied to many times before. but I don't think I ever saw anyone do it as well as this guy. He didn't bat an eye or shift his weight.

I knew better than to even hint at the fact that I knew his fucking name wasn't Ness. He was probably reading my mind while I stood there.

But I quickly realized, he told me A name. I think he liked me.

I liked Ness. I didn't know him, but I liked the way he carried himself.

Someone was running a counterintelligence operation in the Honduran jungle against us and Ness was pissed. Hell, I was pissed.

I was asked to continue taking notes and work with Ness and his partner. Over the next few weeks, I learned that the underlying root cause was that the Sandinistas had placed a priest in the town right in the intersection. He preached anti-American sentiment and told the townspeople in the local villages that the United States Army was building a road to invade Nicaragua.

The road that we were building actually paralleled the border with Nicaragua. Why on earth would we build a road TO Nicaragua. It was absurd, but the townspeople believed it.

They hated us for this. They thought we were there to use them and to invade Nicaragua and put them all in danger. The Sandinistas built and leveraged the traditional Latin belief that the reason that they are so poor is because America is so rich. In Central America many folks believed that we somehow sucked the wealth out of Central America.

This animosity is referred to as Neo-colonialism and it colors and informs much of our relationships in Latin America.

Word spread about my map and note taking. The senior NCOs were all over me.

"YO ROUX! "You did this on your own?"

I took it all in stride.

One of the Sergeants handed me back the papers he got from the civilians and gave me the 2nd best piece of advice I had gotten to that point.

"Next time keep your pages in your notebook. Never tear them out."

That simple piece of knowledge would carry me for the rest of my life and help me achieve the successes that were coming. But I didn't know that then. I just nodded and gave the obligatory "Roger that Sergeant."

I learned the Priest was "dispatched" from his post; At least that was the language they used to tell me that he would never be seen or heard from again. Within a week the Catholic diocese installed a new priest into the region. That priest was very pro-American. And he had a green-colored notepad just like mine...

Chapter 7.   Navigating Local Dynamics

My father's relationship with Argentina was more valuable than I knew. She had very powerful friends in the Honduran Military and Government. The Army wanted to leverage all of the relationships that my family had. I began working in town on behalf of the Sgt. Major and my new civilian friends. They would let me leave the military barracks and go to La Ceiba whenever I wanted.

All of the rest of the soldiers didn't understand why I got such good treatment. That I could just leave on a private motorcycle whenever I wanted and drive 5 miles to town and spend the night there, no questions asked.

Soon thereafter the Sgt. Major started to ask me to help with some of the unit's needs: more soldiers would come to the base, and we would always need refrigerators, window-installed air conditioners, pots, pans, linens, beds, cots, whatever. There was no end to the consumption needs and the exponential growth of our unit in Honduras or for our other friends in the field.

I was asked to go to the city and trade to obtain the things that we needed. I was dumbstruck. I did not know how to trade anything, nor what I had to barter with. So I asked the Sgt. major what I had available to trade and he walked me over to the fenced-in area and said "Take anything in here and trade it for what we ask you for. Ask the Team Sergeants what they need and flex as necessary. Just keep good notes."

Roger that Sergeant Major.

I walked into the compound with thousands of gallons of military spec oil, hundreds and hundreds of tires, and all manner of military supplies. I took an inventory like Bobby Sparks taught me, and put a "No-Entry" sign on the gate, changed the lock and began collecting a list of what was needed, and I set off to town.

At first, the local folks couldn't believe it. This young kid that they knew was out last Saturday with Argentina, was now in a United States Army uniform asking for refrigerators and then more refrigerators. They wouldn't believe it until she sat down with me and the city's elders one day.

"You can trust him.," she said. Then she got up and left.

They all looked at me. Waiting. The Old Honduran man who was doubling as my Spanish tutor finally spoke.

"Write down everything you want.

I jotted down my list and showed it to him. Not even considering tearing out the pages.

"Which of these will you take in a trade, because I don't have money."

Everyone was shocked. It was true. I hadn't been given money. I was told to trade and barter not buy.

One of the elders leaned forward and read my list. A big smile swept over him as he looked up at me.

"Is that motor oil?"

Over the next 60 days, we had just about everything we ever wanted and the locals got everything they needed. I loved the business and I started to be treated even better by the locals. Everywhere I went in town, everything we touched, we were the golden boys. The locals knew that they were always going to get a good deal from me. They trusted me and vice versa. We brought profitable business to their town, and they loved us for it.

I became so popular and ingratiated with the locals, that I was soon tasked with acting as the "go-between" between the 7th Special Forces Group and Local people. If there were supplies that we needed and didn't have, I'd go and negotiate with the Locals to get them by trading some of our supplies.

They needed 2,500 gallons of Mil-Spec oil, 500 tires, and 6 trucks, while we needed refrigerators air conditioners, and furniture. It was a win-win for everyone.

I got really good at this. The Locals would see me and say "There's the Soldier." I was like part Santa Claus and part Mr. Tambourine Man except I wasn't selling drugs. In fact, I wasn't selling anything, I was a trader.

A Blackhawk pilot once asked me "Can we get 10 refrigerators by Friday."

It was Wednesday.

"No problem." I smiled at him and had them for him on Thursday for good measure.

SFC Sanocki the 7th SFG(A) Flight Medic approached me one day. I always liked him and he always treated me well.

"Yo, Roux. I need 20 small refrigerators."
"I'll get them."

"And throw in some ceiling fans too."

Easy work. It was amazing what we could trade for oil and tires. It was better than cash. I had a guy who owned several gas stations in town and he was always needing gas or oil. Of course, the Black label Scotch and good old-fashioned currency never hurt either. It was all about everyone taking care of everyone.

My reputation as a fair and genuine negotiator grew quickly among the people in La Ceiba. I mastered the local language and nuances. I enjoyed my celebrity, and it only made an already cocky 19-year-old that much more confident. I walked around as if I owned the place because I was making positive things happen, or at least I thought so.

The only thing that slowed me down when was when I had to rotate out.

By rule, unless you were either deployed during combat or otherwise assigned to a particular duty at a Base, the Army could not legally keep you anywhere for longer than 90 days. The 7th Special Forces Group was no different.

It's called TDY – Temporary Duty Station. According to the Army rules, a soldier could not be in a place for longer than 90 days back then before the Army had to give him a different type of orders because you just can't have a soldier stationed at Ft. Bragg, have his family move there, do all of the living things as a family and send him to Honduras and never tell him when he's coming back.

This whole thing forces the infrastructure of an Army unit like a company or battalion to treat their soldiers within reason. They're not allowed to just send you on detail forever to bum scratch wherever. You can send me, but you gotta assign me to the unit that's there because that's my duty station and now you have to pay to have my family go too.

So every 90 days we'd rotate back and have to stay there for 30 days or longer,

The special forces are smaller units and have different teams, so we'd constantly have teams rotating in and out, coordinating with each other and keeping the mission moving seamlessly.

But a guy like me had a different role. I would go a lot more often and I wouldn't stay for 90 days. I would go back and forth and back and forth. They put me on civilian aircraft. Nobody did that.

You got to break up the monotony and for those guys that may have been homesick, you knew that your time to leave was always just around the corner.

Once during our rotation out, we were back at Ft. Bragg and one of our commanding officers gathered everyone.

"I need two volunteers for some really difficult detail."

My hand was up before he finished his sentence and Ronnie Philpott with the half-bitten ear was hot on my heels.

"I figured it would be you two."

Ronnie and I both thought we were hot shit and we never minded a challenge. We had no idea what we volunteered for but whatever it was, we were game.

During that time there was a strong sentiment of unity and civic pride that was sweeping through the country. The military was no different. The Chairmen of the Joint Chiefs of Staff designed this thing called the Army 100. It was a marathon canoe race of 100 nautical miles from the steps of the Pentagon to the Chesapeake Bay!

I was excited and anxious to do it. Ronnie's enthusiasm matched my own. We only had a short time to train, but we were going to be ready and we had every intention of winning this race, because that's what we expected; to always win.

We were picked up from Pope Air force Base, and flown to Dulles. From there we were picked up by helicopter and flown to the Pentagon. Not near the Pentagon, AT the Pentagon. General Wickham came out and greeted us and told us our chopper was too small and we would all need to take his. We walked to his Helipad and climbed into his helicopter with him. He was stately and commanding without even trying to be. Still, he greeted us and made it clear that we were special for being a part of this monumental event. It was a surreal moment for me, but I kept calm and wondered how proud my Mother would be if she could see me at the moment.

This wasn't a regular helicopter. It was huge like the Sikorsky VH-3D Sea King Marine One that the President flew in. General Wickham took us to the White House and we hovered 50 yards above it! I could've literally thrown a rock and hit it we were so close. This was my first time seeing The White House in person and it was a view that most would never have the chance to experience.

We flew back around to the Washington Monument and we hovered so close to it I could nearly touch the top of it. General Wickham and his staff on board, seemed to marvel at the monument for a moment. Then he turned to us.

"We're going to take you on a reconnaissance of your 100-mile canoe course."

"Yes sir."

The chopper then flew above the Potomac River for what seemed like an eternity. It's one thing to hear the words "100 miles": it's another to actually fly in a helicopter for 100 nautical miles down the Potomac. The sheer size of the river is daunting. Considering we were about to canoe down the river for 100 nautical miles, I felt like an ant in a windstorm.

I loved every minute of it.

The race was a grueling 3-day race with 3-foot waves in an open Grumman canoe. Ronnie and I approached it like we did everything else; with zero fear and a doggedness to succeed. We tried, but it wasn't our time. We finished in 6th place, out of 6 teams. I wanted to win, but the whole process was so rich and rewarding that it outweighed the result.

When I arrived back in Honduras, the locals were happy to see me. I made a few quick trips to visit Dad and Argentina. Dad was living the life and he and Argentina were happy together. There were rumblings amongst her family and friends still that Gringo Joe was married. I wondered that myself. I knew he and my Mom were divorced. The paper trail, devastation, and the attorneys made it obvious.

So no, Dad wasn't married to my Mom.

He was married to Lynn Decker.

Lynn was great and always cared about us. She was jovial, and intelligent and knew what she was getting with Dad. She treated us kids with respect, and we treated her poorly. It's a shame that youth cannot understand the impact we have on adults. Lynn deserved our respect as Joe's children, and we carried our unacknowledged anger on our sleeves.

Dad told me he and Lyn divorced, so I believed him. I had no reason to think he would misrepresent that truth.

Plus I couldn't be too concerned with my Dad's personal life. I was busily trying to build a life of my own.

Chapter 8. I Can't Even Swim

When you're successful at something, people sometimes believe you can do anything. Be careful of your own success.

The following week I was asked by a team Sergeant named Bracewell to assist his team on a hot mission. Well, he didn't really ask but rather told me I was going to help.

"You're going to accompany us to our POI (Point of Insertion)." SFC Bracewell looked at me and I had nothing for him.

"Yes Sergeant."

"If we need you as an aux operator, all you gotta do is hold onto this box, careful they're detonators and hold onto or near the chem-light on my ankle during the dive. If we don't need you, you'll pull security and wait with the medics."

Dive?

I looked blankly at SFC Bracewell.

SFC Bracewell knew that for a short time, I worked at the 7th Group's Scuba Locker on Son Tay Boulevard for Master Sergeant Wishart. Wishart was a giant man. "TOP," as everyone called him, was one of the most respected soldiers I ever met. He wrote most of the military diving manuals. He taught me how to repair and maintain Drager re-breather gear at the Scuba locker. This gear removed all bubbles from exiting your regulator which prevented any sign of your dive on the surface.

That was the extent of my diving experience.

Bracewell didn't give a shit.

Bracewell handed me my kit with the Drager CCUBA (closed circuit underwater breathing apparatus) Rebreather.

Someone handed me a small box of explosives as the Chinook took off from our SCIF, we had a Zodiac slung underneath the chopper. We fast roped down into the boat and the crew chief cut us loose from the chinook. Waves rocked our Zodiac boat back and forth.

Rebreathers are an incredible piece of equipment. It recycles the air that you breathe when scuba diving, filters out carbon dioxide, and uses a small bottle of oxygen to fortify the air you are breathing. I didn't know shit about this equipment except to follow the repair manual, to identify defects, and order parts.

But that was way more than I knew about the detonators I was holding.

And I couldn't swim. At all.

"Roux, you're going with us." Bracewell had spoken to someone on the radio and the decision was made.

It wasn't a question and I wasn't meant to answer or ask for any more details. And I didn't.

I couldn't swim.

But here I was, detailed without notice to support a mission I knew nothing about and nobody ever asked me if I ever went scuba diving. The first dive of my life was a mission to place underwater mines in the Bay.

One of the guys with us seemed like a civilian and was talking about how the year before they were mining the harbor and Congress didn't like it and put a stop to it.

"Then What the fuck am I doing here?"

We finally reached the dive site. I listened to Bracewell give me some simple instructions on my buoyancy vest and we dove into the abyss. I latched hold of his ankle as we descended and kept a panicked focus on his blue chem light, trying to keep my depth level with his as we began to drift along the reef outside the harbor. Our dive took us 80 feet below the surface.

80 fucking feet!

In my first dive ever, completely untrained and a part of a distinguished group of soldiers who prided themselves on training and preparation, I had just dove 80 fucking feet!

I was scared shit-less and prayed it wouldn't be my last time. I carried the detonators as asked, but they never asked for them nor used them. I wondered on the way back to base if the whole thing was some sort of test. Was there a Colonel or even a General somewhere who wanted to see what I was made of? Were they trying to test me for something bigger? Some other mission. Probably not. The Army itself was a test. If you failed, you got sent home. If you passed you got promoted and advanced.

Sometimes all you need to do is show up, shut up and put up.

This was my first and last scuba mission ever.

Chapter 9.   The Missing Pallet

Though I didn't want to leave, I had to go back to Ft. Bragg sometimes. On most of those trips, I'd drive up to New York to see my mother, and she would listen intently to my stories. I knew she was proud, and it made me feel good to know this. But duty always called me back to Honduras.

One day I returned to Honduras and a Captain, with no name tag, entered the bunker at JTF Bravo SOC at Palmerola AFB. I had never seen him before, and he looked serious. He scanned the room and noticed me. Then quickly approached. He showed me his I.D.

"I'm with U.S. Army CID."

"Yes sir. Captain," I said. Unsure of why someone from the Criminal Investigation Division would be asking for me.

"You are under arrest Corporal."

I was shocked. I turned to the Sgt. Major.

"Is he talking to me?"

"Yes Corporal. You have to go with him."

I couldn't believe it. I had never done anything in the Army or as a civilian to get me arrested. I was a rules follower and stickler for doing the right thing.

"Do you have a gun?" The unknown Captain asked me directly.

If I didn't know before, I really knew then, that this was serious. Soldiers don't carry guns. Guns are things that you go hunting with. Soldiers carried weapons. We never used the word "gun." In the Army that was like calling someone by their first name.

I gave the captain my weapon and all my ammunition. He made me take him to my room. There he rummaged through all of my belongings and asked to read all of my notebooks.

I complied and remained respectful, but inside I was really pissed off. I was one of the good guys I thought, and I was scared of what was going to happen to me and why.

He then informed me of my rights and explained that I was under investigation for misappropriation and theft of government property.

What?

When you work for the Command Sergeant Major or frankly any superior noncommissioned officer or officer and you follow your orders it doesn't even occur to you that what you're doing could have been wrong.

During the investigation, I showed him what I was trading and what I received in return. He wanted to know why I was doing this, and I showed him my notebook of every transaction I had. I took him to the places that I traded it with, and I showed him just about every single thing that I was asked to do, who I did it with, why I did it, who told me to do it, and when it happened.

He already knew all about that and didn't care.

This went on for days and then he finally told me that I was being accused of stealing an Air Force pallet of long-range reconnaissance patrol rations (LRRPs), a second Air Force pallet of ammunition and a 500-gallon Blivet of jet fuel. He knew I was trading all kinds of things for the leadership and suggested that they should be in a shit ton of trouble themselves, but he was focused on me.

He said they thought I might have traded materials to support the enemy.

"OH, MY FUCKING GOD!" I thought.

The irony. I had been brought to Honduras to keep tabs on my dad and his friends to help connect the dots for U.S. Intelligence and now I was accused of selling shit to the Sandinistas. No fucking way.

I knew I was innocent. Hell, I've never even seen an Air Force pallet of ammunition and I've never seen an Air Force pallet of LRRP's.

"I didn't do this. I have no idea what happened to those pallets." I looked at the captain earnestly.

He studied me and I could tell he believed me.

"You've got three fucking days to find that fucking pallet."

"Can I have my weapon back Sir to take with me to find the pallet?"

He looked at me. Shook his head.

"Do you think I'm fucking crazy?"

That meant "No."

He had to deploy back to the United States and would be back in five days.

"I'll have a Sergeant go with you," he told me as he left.

I had a really good understanding of how things were being traded and where it could go if someone wanted it. His Sergeant and I took the Toyota Hi-Lux.

My mind was already working. I knew that an Air Force pallet was 12 ft.² and was 2 inches thick aluminum at its base. Empty, it weighed 337 pounds. Those things carried really heavy shit. I asked myself what kind of vehicle would you need to move that?

As I tried to understand where all the pallets went, I got some field manuals and calculated the weights. I realized that you couldn't drive an Air Force pallet on the road. It was too wide, and the jungle roads were unforgiving. So, if someone was going to steal everything on the Air Force pallet, where was the actual Air Force pallet?

It couldn't be thrown away. It couldn't be burnt. It couldn't be hidden. It would have to be buried and that wasn't going to be easy either. I didn't think somebody interested in stealing that much stuff would take the time to bury the damn pallet.

It couldn't have just disappeared.

I concluded that the only way for that pallet to move was by air and that would mean a helicopter picked it up. From the moment I joined the Army I read everything I could get my hands on about military equipment and history. I quickly figured out that the only helicopter that could pick up something that big was the Chinook. There were only seven of those in Central America! I just needed to know who the pilots were.

I had a lot of friends that flew Blackhawks with live ammunition. Everyone knew that if you needed something I was the guy to get a hold of in the 7th. The Blackhawk pilots always wanted 50-pound burlap sacks of lobsters. And they got live lobsters because I knew I could get a ride anywhere I needed to go. They used to fly to the beach and land with 55-gallon drums cut in half filled with water and just have a lobster boil with beer.

I sat down with some of them over a beer and learned about the flight traffic patterns, where all the manifests were, who was in charge of the manifests, and where to find those people. I knew that the Forward Area Refueling Points (FARPs) always had a shack and a clipboard.

That clipboard had the tail number of each helicopter, along with its origin and destination. There were only seven FARPs and of these bases, all I needed to know was when the stuff disappeared, and I could track it.

It took me three days, but I found the pallets and the jet fuel.

The ammunition was all gone, the rations were all gone, and the jet fuel container was empty with 7th Special Forces Group spray-painted in red on the rubber. I took pictures of all the shit with my Kodak camera.

I arrived back at Golson Air Force Base without a shower and four days of little to eat and running on combat ration sleep. But it didn't matter. My mission was accomplished. I had done what I needed to do to protect the good name of the Sgt. Major and of course myself.

The CID Captain arrived soon after.

As I began to walk them through the details of my discovery when two guys in civilian clothes walked up to the Sgt. Major with a third person. The third was the Commander of the 7th Special Forces Group. He was a Full bird, Col. God on earth, little glasses, big fat cigar, rarely lit. He was a badass killer.

A minute later I was pulled aside, asked for my notes and told to forget about what had happened.

"Everything's fine. It's over." The Colonel nodded at me as he spoke, browsing through my notes.

I could feel the stress escape my body in one fell swoop. Once again through hard work and diligence, I managed to strengthen the reputation I worked so hard to build.

The CID Captain walked over to me with my M-16 over his shoulder. He took it off and handed it to me.

"Good work Corporal. My work here is done."

That was it. I beamed inside.

As a result of my work, my stock in trade went up even higher. I was promoted to Sergeant.

I felt validated. I was always worried that I wouldn't be able to live up to the level of the other soldiers around me but now I was someone. I felt invincible though there was a time looming in my future where I would learn I wasn't.

Chapter 10. The Briefcase and Handcuffs

Despite all the successes and accolades that I achieved, I was still unsure of what my future in the Army would look like.

I had applied to the United States Military Academy Prep School, (US-MAPS) twice and been denied both times, so I was beginning to wonder just how high I could climb through the ranks. I had the aptitude and the knowledge to be a West Point Cadet, but I simply hadn't arrived in the Army with the best grades and academic history. I was 20 years old and 21 was the age limit for incoming West Point soldiers.

I knew that I had to try again. I had come to Honduras and worked in Intelligence, thwarted a potential uprising in the jungle, went on a dive to plant mines in the Gulf of Honduras in the middle of the night, and busted my ass to make my mother proud. I was determined to give it another shot.

This would be my final try. I wrote this on my application and emphasized that "if I didn't get in this time, I would age out of being able to apply."

There was nothing else left that I could do.

My work in the jungle had spread quickly among the officers in Honduras. I had proven that I was capable of anything that was asked of me and then some.

One day Command Sgt. Major "Peg-Leg" Pete Parker approached me.

"Corporal I need you to take something to Mocoron right now."

He was the highest-ranking enlisted man in the 7th Group. I didn't hesitate.

"Yes Sergeant Major."

I answered but no idea on earth how I was going to get to Mocoron. It was an impenetrable trip, and NO transportation went directly there. Let alone one with a daily schedule.

He handed me a briefcase and a pair of handcuffs and said, "I need you to deliver this by tomorrow morning."

I waited because surely there had to be some other instructions.

"You have my authority and the Colonel's authority to tell anybody that you speak in the name of Colonel Waghelstein. You tell them the Colonel said they need to do whatever they need to do under the Colonel's authority to get you wherever you need to go."

With that, he turned and walked away.

Mocoron? Are you kidding me?

I was in Goloson AFB in La Ceiba. There was a truck leaving for Palmerola AFB and I knew I could get anywhere from there. Planes left there and flew all over Honduras and Central America. I hopped on the truck with no real idea of how the next part of my journey would unfold.

6 hours later, I got there, and they couldn't do anything for me.

No aircraft were available and anything else would've taken too long. I headed to this Forward Area Refueling Point (FARP) because I knew there were helicopters there. They always had Chinook helicopters there.

I arrived there and spoke to the Chief Warrant Officer.

"I need a ride to Palmerola, right now sir."

I showed him a map. The guy looked at me and noticed the briefcase. His face said, "Whoa, who the fuck wears a briefcase and a set of handcuffs AND a uniform?"

"Well, I can only get you as far as here. No fuel to make it all the way to Palmerola and they may not have fuel at the FARP when we land. They don't always have fuel when we need it there." He pointed to an isolated spot on the map, about an hour outside the airbase. It wasn't where I needed to go.

"But you're not on the manifest, so I don't know how you get out there."

Manifest?

"I have orders from the Command Sergeant Major to deliver this briefcase by tomorrow morning."

He looked at the handcuffs again. I knew he was sizing me up.

I said, "Look, get your radio. Here is the frequency for the 7th Group and you now have orders to take me there, right now sir and I'm sorry if this seems disrespectful, and I know I just a corporal, but we need to go right the fuck now sir."

I cringed inside. But he knew the same thing that I knew: No one lies in the Army and if you do, you're really stupid because you're going to jail. He looked at the briefcase again.

"What's in the briefcase?"

"I don't know, and I don't care," I answered.

"It's in the fucking opposite direction."

"I know, but I have orders."

The guy kicked the ground and cursed at no one in particular.

He faced me with a resolved look.

"Roger that."

He picked up the radio and told his people what he had to do. Everyone else on the other end of that radio acted as if the Colonel had just been standing there and changed all of the manifests in the region. It was done.

I climbed into the Chinook. It was loaded with gear and supplies.

He took me as far as he could go and dropped me off at another FARP.

I hopped in a deuce and a half and took off to Palmerola AFB.

It was hot as shit there and the concrete tarmac could melt the rubber off your boots.

I walked to the flight line, found the pathfinder shack, and asked when the next C-130 would be headed to Puerto Lempira and waited.

I ended up waiting for 6 hours, it felt like forever.

I heard the C130 Hercules with its 4 prop engines approaching long before I could see it. As it touched the ground, dust and dirt flew everywhere.

The pilot throttled down the engines and the plane slowly cruised down the runway and turned around, ready to take off again. The Men immediately began to refuel the plane. I approached one of the pilots and said, "I need you to take me to Puerto Lempira immediately."

He looked at me and laughed.

"Sure."

His sarcasm matched my doggedness.

I held up my arm to show the briefcase chained to my wrist.

"Oh shit."

He knew this was serious. His plans were going to have to change.

"I'll take you, but we need to make a drop along the way."

Drop?

About 30 minutes into the flight, the ramp went down and 6 pallets with parachutes went out the back door.

We landed outside Puerto Lempira about 45 minutes later.

This place had a big ass runway. It was near La Venta. You can't get there by car because it's across the mountain range and it's surrounded by water. I got dropped off and no one was there!

It is fucking desolate. No food. Nothing. I started getting a little nervous because I had no idea how long it would be before I could get out of there. Eventually, I saw these three guys walking around.

"Should be a flight in about three hours." I was happy to hear there was at least some sort of schedule.

I eased into a place in the shade and sat down to gather my thoughts. I wondered what was in the briefcase that was strapped to my wrist. Whatever it was I reasoned, it had to be very important.

Eventually my ride showed up, a Huey helicopter.

I approached the pilot as several people got off the helicopter. Where the fuck were they going? I thought.

"I gotta get to ODA 735."

He knew where I was going and knew enough to know that I was on an important mission.

I climbed in the Huey with 5 other people. We were at the border of Honduras and Nicaragua on the Mosquito Coast. A jungle so thick that as soon as you hacked away vines and a century of dead brush. As soon as you entered the temperature dropped from the high 80's to something in the low 50's.

Hundreds of feet above the canopy was closed to sunlight, and little grew on the jungle floor. Every unknown creature was louder than the fears inside your soul and it was as dark as night, at noon.

It was both unforgiving territory where nature could devour you and hostile because other men left tracks on the jungle floor.

Helicopter exhaust flew everywhere. I was glad that this ride was going to be short. I glanced out the side window at the Nicaraguan countryside.

Suddenly, the pilot screams out, "We've been painted! We've been painted!!"

Everyone ducked down expecting an anti-aircraft missile to rip through the side of the chopper at any moment. The pilot was in full panic, as he dodged the aircraft across the sky, trying to avoid any incoming enemy fire.

Luckily it was nothing.

The guys from the 7th Group had pointed their laser at the helicopter fucking with him. The pilot was pissed, but he laughed. Those lasers would make the missile warning alarm go nuts. In the jungle, guys were always fucking with each other.

We finally landed and I hustled over to a team sergeant who expected me. It was nearly dark, and he took me over to the fire and offered me some chow and told me where to rack out.

SFC Bruce unlocked the case from my wrist then unlocked the case. The team gathered around the fire; I leaned in...

It was the mail!

That evening, I drank a beer and reflected back on my mission as I planned my journey back to La Ceiba. I thought about just how important little things are to people when they don't have much. These soldiers were far away from their families and friends, but those letters and cards were their lifeline back to their respective worlds. I had helped brighten their days and that felt good to me.

The Team in Mocoron loved me for delivering their mail. I made several friends that I grew very close to. They appreciated me. That's what real Brotherhood was. Of course, I didn't know that there was mail in that briefcase, but it didn't matter. The truth is if I had known it was their mail, I would've walked the entire way if I had to.

A couple of months later I got accepted to West Point Prep School.

Chapter 11. The First Time I Almost Died

I rotated back to Bragg for two weeks and returned to Honduras and had a chance to say goodbye to my dad and Argentina. My Dad was proud that I was going into West Point Prep, and he made it a point to tell everyone.

"My son's going to West Point!"

Most of those people had no clue what West Point was, but it didn't matter to him. Even when I corrected him about it being West Point Prep he didn't care. I think in some sort of way he felt like he was responsible for finally helping one of his children succeed. Perhaps the reason I was turned down for two years was that the intel leadership didn't want me to stop providing data. I'll never know...

Regardless I didn't bother to burst his bubble.

Two days before I was to leave Honduras for the last time (or so I thought), Argentina and my dad threw me a big party with all of their friends. It was a hard night of drinking and good times, and the next morning Argentina cooked me this incredible breakfast with the best bacon, eggs and frijoles fritas anyone could've imagined.

She even made this coffee, crazy enough, from freeze-dried coffee and milk and boiled them in a pot. Amazing. I hugged my dad and kissed her goodbye and hopped on my motorcycle for the relatively short ride back to the base.

I was feeling good. It was a beautiful early summer morning. I was wearing a t-shirt and shorts, feeling the warmth of the early morning sunlight. The wind blew through my hair, and I gunned the 2-stroke motorcycle, to about 40 m.p.h. I approached the bridge that crossed the Rio Tinto toward Goloson

Air Force Base. The river was 30 feet below and was normally pretty deep because of the runoff from the mountains. There was no shoulder on the road as you approached the bridge. If you drove off of it, you would fall the 30 feet into the river and likely drown.

I saw a mule and cart approaching me as I neared the bridge. I slowed just a bit, knowing that I was going to pass them. Suddenly, this car darted from behind the mule cart and headed right for me. It was going to be a head-on collision.

I had to make a quick decision; either I fall into the river which wouldn't have been good or...

... I had to lay the bike down on the gravel dirt road.

It was worse than sandpaper. I felt my skin being shredded into small, bloody pieces as I rolled over several times. I finally skidded to a stop. I could hear the motorcycle's engine still running.

The car never stopped to check on me.

I dragged myself towards the motorcycle, hoping that no one from the Base had seen me. We were supposed to be leaving to go back to the States and I was going to be in deep shit about this.

One of the guys from my Unit saw the whole thing.

"Roux you okay." He had this look on his face that made me know I was in bad shape. The truth is I didn't know if I was okay, but I did have the wherewithal to let him know what I was thinking.

"You can't tell anybody about this."

"Are you fucking kidding me. You're in bad shape!"

I managed to prop the bike up and climbed on. I drove away slowly. Dazed. In shock and bloodied. No matter what though, I couldn't go to base like that.

"Adam! Oh my God! This is bad! This is bad!"

The moment Argentina saw me she began to cry. My Dad tried to calm her and me though I was still in shock and didn't understand the severity of my injuries.

"This is so bad!" Argentina couldn't stop.

She eventually settled down and started to administer First Aid to me. Each time she added ointment to my wound, she started to cry again, only louder.

30 minutes later a truck pulled up to the front of Argentina's Pharmacy. The guy from my Unit hopped out with another guy I didn't recognize.

"Roux."

They helped me into the back of the truck and drove off. Argentina was still crying as we pulled away.

All I could think about was the fact that we were supposed to leave the next day. I was in deep shit.

When I got back to Goloson Air Force Base the guys put me on a cot. I tried to hide my injuries by covering up my bandages with my uniform shirt and pants. They went and grabbed SFC Sanocki. Sgt. First Class Walter Sanocki was the 18D medic for my unit.

Sanocki saw me and his eyes got wide like saucers.

"Holy Fuck. I've got you man. Don't worry."

I couldn't worry. I was in too much pain.

Sanocki said my wounds needed disinfecting. But we had already packed all of our shit onto these 12 x 12 8-foot high Air Force pallets and had them laid out on the tarmac ready to be loaded into a giant C141 Starlifter.

Sanocki would have needed a detail to unpack one of the pallets.

A Special Forces Medic is universally referred to as "Doc." He prescribed drugs. He has all the drugs on the planet.

Sanocki looked at me again.

"Goddammit I palletized everything, Fuck!"

He reached into his side bag and found a little bottle of Percocet. He gave me 3, 500 mg capsules.

"Take these."

At the time, I had no clue what I was taking. When Sanocki gave you meds, you took the meds. You did what the Green Beret medic told you to do.

We were leaving in 4 hours. Wheels up. People were running around and preparing for departure. I was looking at the plane. I sat on a cot with an IV in my arm and I was bandaged up with my shirt and pants over it all so I didn't cause too much attention.

65 people were waiting to get on that plane.

Sanocki handed me a warm 16 ounce Tall-Boy Budweiser.

"You need this."

"Okay."

I wasn't really a beer drinker, but if he said I needed it, I must've needed it.

An hour later we were playing cards between 2 cots and a brown box. 4 guys. Sanocki gave me another beer.

My 2nd beer.

It was another tall boy. I drank it. Immediately I knew something was wrong. I looked at Sanocki.

"Sarge, I'm having trouble breathing."

He looked at me and then I went into anaphylactic shock! I passed out right there on top of the card table.

Captain Ellerbee went bat shit crazy, and rightfully so.

"What is going on!"

He didn't want any unwarranted bad attention and he had earned the unit's respect and we were letting him down. We had a new Colonel now, named Colonel Williamson. He didn't know me, and I didn't know him.

Until that moment.

The Colonel quickly learned about my accident from what I was told. I was fighting to remain conscious in the middle of the base terminal.

Our fucking plane had to leave. Absolutely.

The whole unit had to leave and they could not leave me behind. You can't do that. It's not about, " We never leave a man behind." It means "You are not allowed to leave anyone behind."

I wasn't conscious so someone would've had to stay with me and no one was allowed to stay. It was a real fucking mess.

If we didn't leave, the unit would have had to report that it missed its departure deadline. It's called being Blacklined. That means you are combat ineffective and that means that goes all the way to the top.

"What, caused the 7th Group to fail to depart from Honduras!??"

Someone was going to be in real trouble.

Sanocki tried to revive me to no avail. He rubbed my sternum which was supposed to wake me.

Nothing.

They had to de-palletize all of Sanocki's gear.

He got a needle and gave me an Epinephrine injection in my heart.

"Hey Adam how you doing?"

"Hey Sergeant."

I struggled to focus my eyes on Sanocki.

"Wow I'm really sleepy. What's going on?

60 seconds later, I went down again.

Sanocki couldn't wake me up. He gave me more Epinephrine. This went on for half an hour until he finally stabilized me. That meant that I was ambulatory and could walk, which ultimately meant I could get on the plane and everyone could leave.

Everyone kept asking Sanocki, "What the hell happened? What happened to him? How did that happen?" The other Green Beret medics from other units, who were also getting on the plane came over to see what the commotion was all about.

None of them had ever really liked Sanocki

One of them leaned into me. He knew that something wasn't right. He was so close I could feel his beard.

"Fuck."

He studied my wounds. I could see his mind working. He knew this should not have happened. I had wounds on my knees, hips, hands, arms, and toes. That was road rash. I should not have gone into shock.

"You almost died today."

"Did I?"

"What exactly happened?"

"I don't know Sergeant Sanocki..."

Before I could finish Sanocki rushed over and whispered into my ear.

"Do me a solid Roux. Don't tell these fuckers anything. Please."

I had never heard Sanocki say "please." No one in the Army, especially Green Berets, ever said "please." I knew this was serious and though he had just nearly killed me, I had to protect Sanocki. He was in my unit, and we had to stick together.

The bearded guy frowned and shook his head side to side when Sanocki whispered to me.

"I don't know what happened. I took some meds. I got tired. I was just sitting there and then I blacked out."

I looked at the bearded guy and shrugged. He glared at Sanocki and left.

Later on, Sanocki came back to check on me. "Thank you, Roux. I appreciate you."

I almost blacked out again. He actually said "Please" and "Thank you" in the same day.

Almost dying hadn't been good for me. I knew Captain Ellerbee and the new Lieutenant Colonel both thought I was difficult. Captain Ellerbee especially seemed exasperated by my early morning motorcycle routines. More than that, he didn't appreciate my tendency to operate autonomously at times, taking liberties that were unusual for someone at my rank.

He shook his head as he passed me, and I knew I was in trouble. My West Point Prep acceptance was now in serious jeopardy, but it was out of my control at this point. Captain Ellerbee was the kind of officer enlisted men respected without question. He showed up every day with character, resolve, and an unshakeable commitment to his values. In a way, I admired him for that—it wasn't just his authority but the example he set that made people look up to him.

Luckily, we left on time.

Captain Ellerbee and I didn't have a poor relationship by any means. I respected him immensely, but I came to realize that, despite my efforts to show deference and respect, he didn't seem to see me in the highest light. Looking back, I understand why. I was still young and perhaps too naive to grasp the depth of the path he'd taken. Graduating from West Point in 1980 and building his career step by step, he had earned every accomplishment. In contrast, I had found myself fast-tracked to a position with the Green Berets—a privilege that, I now realize, must have appeared unearned in his eyes.

Understanding this drove me to prove myself every day. Captain Ellerbee represented the best of America's meritocracy, and being in his presence only strengthened my resolve to earn my place.

I woke up in the hospital back at Ft. Bragg. They had taken me straight from the plane to the hospital. The nurses and doctors gave me nitrous oxide because they had to scrub my wounds which had already started to heal over. That wasn't a good thing because I had dirt and bacteria in them from the long flight back and the accident itself. The pain was intense.

I started huffing this stuff trying to ease the pain, but it didn't work.

A Doctor passed the room while the Nurse was scrubbing my wounds. He saw me huffing on that nitrous oxide and stepped inside.

"Hey man, you can't keep huffing on that stuff."

I couldn't even answer.

The Nurse who was scraping my knee made eye contact with the Doc who shook his head. And left.

I noticed three Generals walk in wearing their white coats. When I saw them, I instantly thought that was strange. Seeing one General is not common. Seeing 2 is only during a huge chain of command change. Seeing 3 is like seeing an alien. I wondered what was going on.

Then I saw it.

They had an X Ray. I was still in the ER triage room and there was an X-Ray and this poor guy had gotten hit sideways with the breach of an artillery gun and crushed his entire jaw off just underneath his nose.

He lost all his teeth and everything was pulverized. He was alive and surprisingly he was fine because it was so violent and so fast there wasn't a concussion to his brain.

I felt lucky. I knew that in a few days, I would heal and be gone.

It took me 2 weeks to recover and the whole time all I could think about was losing my West Point Prep acceptance. And for good reason.

Ellerby had gotten over the incident, even though nothing came from it. Still, it was an embarrassment and reflection of his command because it happened under his watch. He was pissed, and it's safe to assume, Colonel Williams, his Superior was pissed too. They had decided that they were not going to send me to West Point Prep. I would not learn about any of this until much later. I was less than 90 days from leaving and it was about to all get taken away.

But Les Thatcher intervened. He was privy to the work that I had done to help him and his operators. By this time, he was the G2 in charge of intelligence operations for all of the Special Forces in the U.S. Military. The Green Berets, Navy Seals, Marine Recon, Army Rangers; you name it. He was powerful and an unmatched ally to have on your side. Les Thatcher called Colonel Williams and gave him the instructions to let me go.

I hated lying around with nothing to do. I had worked to show that I wasn't a "soup sandwich." I was not lazy and lying there made me feel inadequate. My unit needed me because I was the Armorer and there's only one. I had the special keys. I had the special codes. I had the special room. This weighed heavily on my mind.

My friends would drop by and see me once in a while, but Sgt Bobby Sparks was really the only guy who came consistently.

"I need to get back to my unit." I said it every time he came.

"Adam, I know you. I know who you are. Get better."

Sparks would never go on deployment. He was the Supply Sergeant. Everybody that went on deployment was cool, but there were hundreds of people who wouldn't go. Deployments were only for the guys who were assigned to a team or had a special reason to go.

I had been fortunate enough to have been placed on the Advance Team. The Advance Team are the people who go before anyone else does. They got everything set up. Set up all the intelligence and then they came back and told everybody what the deal was. The guys on the Teams respected me. Most didn't know why I was there, but some did. Sparks knew this. He knew that the Teams at least respected me as part of the process and knew that I was going to come back.

During my hospital stay, Sparks eventually had to get a new Armorer. I felt relieved that my Unit would have someone but saddened that I wouldn't be there anymore. It was okay though. It only made sense.

I got word that my appointment to West Point Prep was still standing. I was ecstatic. The only thing though was that when I was discharged from the hospital, I didn't have anything to do. I was in a flux. The Army was unsure of what to do with me. I was leaving for Ft. Monmouth in 90 days. Should I be deployed again with such a short window? I wasn't the Armorer anymore so that duty was gone. I wasn't sure what to do with myself either.

Until I got a call one day.

"Adam." I knew that voice on the other end very well. It was Les Thatcher.

"Yes Sir."

"I need you to meet me in Honduras."

Chapter 12. The Interview

When I finally arrived back in Honduras, I was invited to go and stay at this guy's house by Les Thatcher. I had no clue who the guy was or what he did, but all that I know is he lived in the middle of La Ceiba and he had this beautiful home. He wore civilian clothes, and I wasn't sure what his rank or business was.

Hell I didn't even know if he was in the military.

"Adam, congratulations on your West Point Prep appointment." Les smiled genuinely at me and the guy offered similar congratulations when I first arrived.

"Adam good to meet you. Les has told me good things about you. You have a bright future ahead of you."

"Thank you Sir." I decided to assume he was just as high ranking as Les and he didn't correct me so my intuition seemed right.

He and Les were clearly great friends and he treated me as if I were his family. I wasn't aware of this at the time, but this wasn't just a friendly visit. I wasn't there because I had just gotten into West Point Prep.

Les introduced me to another guy named Billy. Billy was handsome, 5'9, 190 pounds of solid rock, with a stiff jaw. Billy was a killer.

I was there for an interview and I didn't even know it.

That's how intelligence works. You're not supposed to know when you're the Subject. I was too young and naïve, perhaps even immature to realize that Les

Thatcher didn't just call soldiers up and tell them to come and spend some time with him and his intelligence buddies.

I was still riding the high of my acceptance to USMAPS, I had visions of becoming a Commissioned Officer which can be tough for guys like me who couldn't go to college.

We talked and the guy suddenly got excited.

"Adam, do you want to see my Uzi?"

Uzi? Hell yeah, I wanted to see his Uzi. I had never seen one in my life and I was an Armorer.

"Why the fuck did this guy have one?"

Everyone in the military doesn't just walk around with submachine guns strapped across their shoulders every day. In 1985 at the 7th group, the only subs we had were Heckler and Koch MP5-K's, The full autos we had were AK 47's in three models.

"Sure." I tried to downplay my excitement, but inside I was bursting. Truthfully, I felt intimidated really because this guy was some sort of a big deal. I tried to cover that up though with some machismo that I thought I was supposed to display. Youth.

He showed us his Uzi and I acted as if it were no big deal. I intentionally kept focusing on other things.

He didn't seem to pay much attention and went about his business of showing it to me.

I left the next day, ready to get back with my guys and get one more step closer to West Point Prep.

I never heard anything about this from Les. It would be decades before I even discussed what had happened. I was being interviewed the whole time, I eventually surmised. This guy had some sort of assignment that he was considering having me do and he wanted to see what I was like away from the base. I can only guess that Les had highly recommended me for whatever it was.

Chapter 13. The Path To West Point

I reported to West Point Prep school at Ft. Monmouth, New Jersey in August of 1986. I was confident and ready for this next phase of my life.

Ft. Monmouth was a former military post that was surrounded by the towns of Eaton town, Tinton Falls, and Oceanport, New Jersey. It was about 5 miles from the Atlantic Ocean and if the wind blew just right, you could smell the salty water. The locals loved the soldiers and were patriotic and supportive of everyone at the Prep School.

At that time, Les Thatcher had moved his Intelligence unit from Vint Hill Farms Station to the 518th Intelligence Battalion, a signals intelligence unit, at Ft. Monmouth. Les had an important mission to fulfill at Ft. Monmouth but spent a lot of time outside of Manassas, Va. In Vint Hill. Les was an intel icon and his leadership and patriotism bonded all soldiers that worked for him. For those that got close to him, we became family, knowing we had one single touchstone. Les was a Touchstone for those of us he deployed into harm's way and worse.

Though one of the U.S.'s Intelligence Posts was at Ft. Monmouth, it didn't appear to make sense for Les to be there. There were tons of satellite dishes, very few people, and a lot of locked doors. Why Les was there, I knew I shouldn't ask. Intelligence gathering had to be done on domestic soil as well.

The Prep School itself was a mix of soldiers and mostly high school kids. West Point Prep provided an opportunity to matriculate people with a particular set of skills that wouldn't otherwise qualify educationally for West Point. These guys were great soldiers, who like me, didn't do well in English and

Math in high school. But if you stuck it out the Prep School would allow you to go to West Point and there you'd be set for life.

The Prep School gave you a small dose of Army life. For the high school kids, it was a terrifying experience that left many of them wondering if they were going to make it.

For regular army soldiers like myself, we were generally of the mindset: "Is this a joke? why are you putting us through this mental bullshit?"

The Instructors loved making you do push-ups and sit-ups while they yelled and screamed in your face. They were good at it but they generally lacked that inner killer that comes from having fought in Vietnam where you have had to watch your buddies get murdered in foreign jungles.

They weren't the Drill Sergeants from Ft. Benning.

After a day or so of this, I had just about had enough.

"What the fuck are you doing Rousselle!" The young Sergeant yelled in my face.

I turned and looked at him and took my time speaking.

"Look Sergeant, this doesn't affect me. I'll do everything that you say, but don't bark at me."

"Oh yeah. Oh yeah!" He was trying to maintain his image and tough guy persona.

This was one of those paradoxical situations about the Army. I arrived as a Staff Sergeant with more experience and he was a Sergeant, one rank lower. But that didn't matter. If you're not the officer in charge, you're like everyone else. There are times when a Captain may be in charge of a Major and the Major refers to him as "Sir," because the Captain may be Cadre and in charge of the training. When shit happens no one ever forgets that the Major is the decision maker, but in training those are the rules This has happened to everyone at some point in their military life, so soldiers tend to be respectful on both ends.

But he didn't get it. He was on tilt 24/7/365 because it made him or something.

He kept screaming and yelling and it started to piss me off. He figured he was going to make an example out of me because I was the guy with the most medals and accolades. He knew who I was. I was probably the 2nd most experienced soldier there.

There was one other guy who was Sergeant First Class named Norman Matzke. He was a soldier's soldier. I even looked up to him.

They took my Green Beret off my head and threw it on the ground.

"None of that fucking matters now Rousselle!"

I had never been insubordinate in my life, and I wasn't about to be to this little young sergeant either so I just smiled or looked ahead. I had already made up my mind how I would handle him.

"You're not them anymore, you're with us. It doesn't matter what you do and we're in charge. And all of you kids are going to learn. Nothing going to save you!"

He made me start doing push-ups while he yelled at me. His buddies got excited about all of this.

I looked around at all of the kids around me. They looked scared. They looked at me and I could almost see their thoughts on in their expressions.

"If they can do that to him, what are they going to do to me?

I smiled and looked at all of them.

"Guys," I spoke directly and firmly to the young soldiers in my class. "All he's doing is, he's going to keep making me do these push-ups until I have muscle failure. Do you understand? And as I reached muscle failure I continued" "Now I'm at muscle failure and I can't do another push-up and there's nothing that he can do to me. The whole thing is over. He's screaming, but it doesn't matter."

That young Sergeant could've shitted a brick." His face flushed with anger.

"Turn over. Do flutter kicks"

I politely turned over and spoke loudly so that all could hear me. "I'm going to do flutter kicks until I have muscle failure. I will do everything this Sergeant tells me to do with a smile on my face."

I flashed a smile to them and I could see some of them fight back a giggle.

"As long as I do that no harm will come to me or you. Ever. Pay attention to what people tell you to do and do it and you'll pass with flying colors. Pay attention because the true lessons are how to listen and act quickly.

The Sergeant was pissed because he knew I was right. I had let the cat out of the bag and there was no going back.

From that moment on, I had a lot of friends. I'd tell them tidbits of advice.

"It's about your mind. It's not about your body." "There's no rule that you have to do 100 push-ups. You do them until you can't. What is he going to do kick you in the face? Nope. They can't do that."

At least not there at West Point Prep. Ft. Bragg was another story.

I didn't mind sharing my experiences with the other soldiers; in fact, I relished it. It was good not to always be the youngest person.

Chapter 14. Bon Jovi Changed My Life

I drove a 1985 Blue Nissan truck. Back then that was magic. I brought it when I arrived at the 7th Group (the car my Dad had bought for me in high school had long been repossessed), after I discovered that soldiers had great credit to buy cars or anything else we wanted. I loved that truck.

One weekend Scotty Haskell in my unit and I decided we would go to a 38 Special concert. My buddy Tim was from the area and wanted to go as well. We had the weekend off so we drove to Massachusetts blasting 38 Special the whole way.

The folks I met were some extremely nice people who had done well for themselves. Everywhere we went folks were incredibly kind and friendly and some welcomed me literally with open arms and a hug. We were soldiers, but we were like rock stars.

We scored the tickets, and we were all excited to check out our band. Some band named Bon Jovi was opening up for 38 Special. BON (who the fuck are they) JOVI, imagine.

Our seats were awesome. We managed to get close enough that we could see the sweat rolling down Jon Bon Jovi's neck when he sang. The place was electric! Their set was unbelievable! By the time they finished, everyone was like "Holy Shit!"

38 Special rocked the place too. They lived up to their top billing, but when we left all you could hear people talking about was, "Did you see fucking Bon

Jovi? Who the fuck were they?" Bon Jovi had wasted 38 Special. There was no denying it.

We passed out sometime in the pre-dawn hours, still buzzing from Bon Jovi and the beers we drank at Tim's place. We woke up late, grabbed some food, and planned another night out. More bars. More beer. More conversations.

The next morning we jumped into my truck to head back to Fort Monmouth. We had one rule: The guy in the bed of the truck can't drink, the guy driving can't drink, but the passenger can drink.

The logic of this was deeply rooted in the minds of 20-year-old soldiers who didn't have a real worry in the world; except to get back to Ft. Monmouth before the weekend leave was over.

So every 2 hours we switched drivers, while the truck was moving.

The passenger climbed out of the window and jumped into the back of the truck while we drove something over 60 MPH. The Driver had to slide over the bench to the passenger seat while he kept his foot steady on the gas so the others didn't fall off the truck and he kept his left hand on the steering wheel. The guy in the back would then come through the driver's window and take over driving duties.

Just like the pros did it.

Some of the beer we bought with our fake ID's needed a bottle opener and we didn't have one. Scotty was in the passenger seat as we approached a big rig.

"Get next to him." He yelled to me as I saw him climbing out of the seat. What the heck was this crazy mutherfucker about to do?

Tim howled. I eased up next to the 18-wheeler and Scotty motioned to the driver.

Scotty stepped onto the landing deck next to the driver's door of the truck.

The driver smiled and looked at him like "What the heck is wrong with this kid." Scotty was undeterred.

"You got an opener?" Scotty pointed to the top of his beer and the driver got the message.

"Pass it here."

The driver extended his arm for Scotty to pass him the beer. Scotty was fearless (and drunk). He handed the guy the beer. He popped it with his teeth and passed it back to Scotty.

"You got one for the effort?"

"Hell yeah."

I gladly reached back into the cooler, handed the beer to Tim and he threw Scotty a beer. Scotty was still standing on the moving 18-wheeler, but he caught it, threw him a second and Scotty handed the guy two beers. I kept the truck steady and on the road.

It was amazing!

"We need another night." Scotty was serious as he climbed back into the truck. Tim chimed in and it didn't take me long to agree. We noticed a cheap hotel off the side of the road near another small town like every other small town in that part of the country. It didn't have much, but we knew there was a bar, women, and beer.

This wouldn't have been a problem if we weren't supposed to be back at Ft. Monmouth.

But we made our decision and stayed. What could they do to us?

We got up early the next morning and left. It felt like it was worth it. The moment we hit the Jersey Turnpike it began to rain. And sure enough, I was in the back of the truck. I pulled up my West Point Prep hoodie sweatshirt around my head and watched the rain falling on the bed of the truck where I sat.

Suddenly this car pulled up next to us.

This gorgeous redhead rolled down her window and looked at me.

"Hey."

I had to do a double take. Was she talking to me? I wasn't sure what to say.

"What Cha Dune?" She had a thick Jersey accent and this incredible smile.

"Getting wet." It was the truth. I was soaked, but even if I could, I wouldn't have climbed back in the cab of the truck and missed this opportunity.

She was stunning. She had beautiful red hair and luminescent green eyes. She was the most beautiful woman I had ever seen in my life, and she wouldn't stop flashing that infectious smile. "This gorgeous woman just appeared out of nowhere and she is talking to me!"

We proceeded to talk for 5 minutes with Scotty holding us steady. I didn't know what to say, but I was just myself. She seemed to like that.

A car slowed in front of her, and she had to slow down. Then two cars cut her off. I banged on the back of the cab to get Scotty to slow down, but it was too late. She disappeared in the traffic.

Scotty and Tim laughed their asses off.

"Dude, you suck!"

"You drove too fast."

"She wasn't trying to talk to you anyway. She was trying to talk to me."

We all laughed at the moment, but I thought, "Damn, I hope I see her again."

Beep. Beep.

The horn was weak, and I barely heard it over Tim's boisterous laughter. I looked up to see this little GMC Gremlin with the Redhead had returned!

"You can't get away from me that easily."

I loved that. She was in pursuit of me.

Scotty tried to intercept her by slowing the truck down, but she wasn't having any of it. She made sure that she stayed where she could talk to only me.

We got to the toll booth and there was a big gap between us. This was the Jersey Turnpike. There were a million cars on the road. She was gone for good, and I had to admit that to myself.

"Your girlfriend's gone Roux." Tim peered his head out of the window.

"Fuck you." I flipped him off and we started laughing again.

20 minutes later it was almost time to change. I was glad because it started to rain again. Hard.

"No fucking way dude!"

I saw Scotty looking out of the rearview mirror.

I turned to see what he was talking about.

That damn Gremlin with my Dream-girl at the wheel was charging like a 3-year-old in a yearling race at Saratoga rounding the 3rd quarter pole!

She caught up to me.

"Where are you going?"

"Ft. Monmouth."

"You want a dry ride?"

"What?" I couldn't hear her.

"Do you want a dry ride?" I heard that.

"Yeah."

"Then pull over."

I tapped on the glass to let Scotty and Tim know.

"Sure." Hell yeah. Did I want a dry ride? I couldn't answer quickly enough.

My friends stared at each other. They couldn't believe it. I had just gotten picked up by a gorgeous woman, while riding in the back of my own truck, soaked, on the Jersey Turnpike.

We pulled over and I jumped out of the truck bed, and she was startled and suddenly seemed afraid. I think she had second thoughts about picking up a soaked soldier on the Jersey Turnpike.

"I can get back in my truck. It's okay. It's my turn to drive anyway."

She looked at me with those deep green eyes. Thought for a moment.

"No. get in."

I happily climbed into her car.

"I'm Catherine."

"I'm Adam. Rousselle. Like you sell." She smiled again.

"Hi Adam."

I was instantly captivated by Catherine. She was more than any woman I could have ever imagined that I would meet in my life. We followed the guys back to Fort Monmouth laughing and talking, as if we had known each other in another life. We just instantly bonded. Love is one of those things that you simply don't know what it is until you've actually experienced it.

It was October 15, 1986.

Two months later, I proposed to Catherine on December 24, 1986 and I made the decision that I would choose her over West Point.

It was the best decision that I ever made in my life.

When we got back to Ft. Monmouth, shit hit the fan.

"I'm sending your fucking ass to Korea. You're outta here!" The LTC. Looked me square in my eyes and I did my best not to make eye contact with him.

"You'll never go to West Point. You think you can just show up whenever you want to?"

"No sir."

"Then what the fuck happened?"

There was no way I was going to stand there and tell him; we got drunk and decided to keep partying. I wouldn't have stood a chance. Tim and Scotty were next. I had to do something.

"My truck broke down."

"Bullshit."

I wasn't sure if I should answer.

The LTC eyed me. He knew that I was an asset to West Point Prep. I had experience and a calming effect with the new guys. They didn't need me there, but it was a luxury to have me around.

"Show me the receipt for the repairs."

"I don't have it Sir."

"Why don't you have the goddamn receipt?"

"I can get it."

"You damn right you can get it. You've got one day to get it to me."

I was fucked.

I didn't panic though. I found a phone book and started calling service stations all along the Jersey Turnpike.

"Hey, can I speak to the Army veteran mechanic?"

"What the fuck are you talking about?"

Most of the people hung up. I would've done the same thing. I figured there were hundreds of veterans in these small towns in upstate New York. Surely one of them had a veteran mechanic on staff. And if it's one thing I know, Veteran's like to take care of Army guys.

I called fifty places before I finally found a mechanic who had served in the Army. When I told him, I was at West Point Prep and had been a part of the 7th Group Special Forces, he was even more excited to help.

"Do you think you can send me a receipt for some basic repairs to my truck? Post-marked two days ago?"

"Tell me where to fax it."

Just like that it was done.

When I went back to the LTC's office. He didn't even look at the faxed receipt.

"Don't get into any more shit."

Tim and Scotty couldn't believe it. I had pulled it off.

Two months later, I proposed to Catherine on December 24, 1986, and I made the decision that I would choose her over West Point.

I learned at that moment that I could recognize love, I knew Catherine brought me both strength and peace. I knew that love was a rare commodity and that half-hearted commitments cannot survive the battlefield and would never withstand the requirements of love and marriage. So, I made the best decision of my life to love her completely.

I knew as well that Loyalty doesn't just mean monogamous. I learned that sitting through the worst of times no matter what comes, pays off in the end. I decided to hold on tightly to her love.

Chapter 15. Trouble On The Horizon

"Adam, I really think you're making a big mistake."

I could hear the earnestness in my dad's voice. I had just told him that I had decided not to go back to West Point Prep, and he was trying to convince me otherwise. The problem was that by rule I was not allowed to date Catherine while at West Point. I realized rather quickly that I simply wasn't willing to give her up, no matter the cost.

"She's incredible Dad. I've never met anyone like her." I loved her, and I knew it, and I was going for it. It was obvious to me she was witty, kind, intelligent and she liked me, a lot.

I liked the way she made me feel, I loved her smile and her freckles. Her father was a U.S. Marine who worked at the New York Times. He was a union man; a pressman, and a journeyman there and very, very proud of who he was. He worked every day, to provide for his wife and his three children. He was a great American, the kind of guy who put stakes in the ground, had a white fence around his house, was good to his neighbors, and was trustworthy to the marrow. Her mom was very religious. Much more so than I ever was. She was firm, kind open-minded and she had a family unit that was cohesive and intact.

I felt I could learn from Catherine. I felt that she had a family background that could help us raise a family in the right way. I was in love with her.

Catherine's beauty was radiant, stunning, and bubbly and every time she walked into the room, she blew it up.

Catherine was graced with the gift that's so rare it's hard to explain. I loved and still do so love Catherine for whom she was who she is and who she's become.

She happens to be perhaps one of the most beautiful people you would ever meet, unassuming genuine with an unbelievably bright shiny personality.

I loved Catherine for who she was.

She resurrected me, she healed me, she loved on me. She's put up with my weaknesses and failures and I just can't believe that I've had the blessings of God to have her sit with me all the days of my adult life. If you ever love someone, really love them, the kind you just want to cozy up to and hold really close and tight, that familiar smell makes you want to close your eyes and melt into them that's how I feel when I'm near her. I prey it never fades.

Catherine and I were married on August 22, 1987.

This sparked an intense debate with my father that lasted for days. I had dug in and drew my line in the sand. I loved Catherine with every part of my being.

I was enthralled and consumed by her and she felt the same about me. I spent hours thinking about her. I stopped hanging with my friends and spent all of my time with Catherine. We were inseparable.

I asked myself, "why me?" How could someone so beautiful and kind be interested and in love with me? Eventually these internal questions became outward words.

"God told me that you would be my husband." Catherine's response was so sincere and heart-warming that I melted inside.

I knew from that moment on, she would be my wife. No matter how much anyone else may have tried to convince me otherwise.

Growing up, I had few examples at home of how to love a woman. I began writing poetry to her every day while I was in Prep school. One day I arrived at her home on a Saturday and she opened her mail and read a poem I sent her three days before. As she read my poem her face turned a shade of red that blushing cannot explain. I learned then that honest heartfelt words create romance which releases a powerful and effervescent love that no words can fully embrace.

I decided then I would romance the hell out of her. Always. She deserved the same type of ingenuity though expressions of love, as I spent on my career.

I learned in those early days that contagions, were not the exclusive weapons of illness. They were sand laden wicker baskets, perfumed with innocence, stained by seagrass and bottled in the deepest, quietest parts of our soul. No harm will ever befall you by demonstrating unconditional romantic love.

Les Thatcher definitely wasn't happy about my decision. Perhaps in his mind, he had a vision for what my military career would be like. He likely thought I would be working for him operationally, perhaps as an asset with leadership potential. I don't know. I do know that he was an ardent supporter and mentor to me, and I appreciated him for all of that. I felt more of a need to gain Les' blessing than my own Dad's.

In the end, he recognized that I wasn't going to change my mind.

"What's your plan Adam?" Les was an Intelligence guy. Everything was always about planning.

"I'm going to re-enlist, go to night school, and become a Staff Sergeant." I wasn't exactly sure how I was going to do the school part at the time, but I knew that I needed to.

I had a vision of going to Officer Candidate School and gaining a commission as an Officer in the United States Army. The problem was this just didn't generally happen for enlisted guys like me. You had to either had been a college graduate or risen through the ranks of an ROTC Program. I had to go to college and graduate.

I left West Point Prep school in January of 1987 right after I proposed to Catherine. I quickly did some research and discovered a program that the Army had that would allow Sergeants to go to school full-time while still being on active duty. The whole thing would be paid for by the U.S. Government with zero debt for the soldiers. The Program itself had been created to help Sergeants get college degrees.

I decided to enroll at the University of Tampa. I didn't want to spend four years in college, taking classes that were simply educational audits of things I had already done in the military so I began to research the classes and my applicable experiences that could be applied towards them.

I lobbied the Army and the University of Tampa with a stance that my military experiences should count towards my degree. At the very least if it did it would shorten my time there at the school, thus decreasing the cost for the government.

By the time I was done, I had successfully managed to lobby the school and the Army to give me credit for two-and-a-half years.

Catherine and I managed to get a small one-bedroom apartment off base and though I was "active duty," I rarely had to report to anyone. My job was to be a student and I became the best student I could be.

! I graduated with a B.S. in Business Management in 18 months!

I immediately applied to Officer Candidate School. The Army refused my request citing the fact that the program that I was in was merely to help Sergeants earn degrees, not to use their degrees to gain consideration for OCS.

Essentially the Non-Commission Officer branch argued I got a bargain and now I was trying to break it.

They argued that the contract I signed didn't allow graduates in the program I joined to apply for OCS. I dug deep into the details in a way that would serve me well for the rest of my military and civilian life.

I discovered that that rule had not been instituted within the program when I signed my contract. Thus, there was no rule against it.

Still, the Army wouldn't budge on its stance. I had watched the tenacity of my mother as a young Boy growing up and how she had worked multiple jobs regularly just to help us scrape by. Seeing her doggedness and determination instilled this same sort of attitude. I refused to be denied.

For four months I fought the Army alone to allow me to go to OCS and eventually, I won. I had done nothing wrong, and my initial intent was not to circumvent a system. I simply had done what the rules said I could do when I signed up for the program. This of course pissed some people off, but what could they do, make me do push-ups? I just smiled and waved.

I had jumped out of the window once again and done everything that I could to reach my goal.

I got orders to go to Ft. Benning, Georgia for OCS, and Catherine and I got a small apartment across the river in Phoenix, Alabama. Up until this point I had been free to come and go as I pleased. This wasn't the case with OCS. It was an intense and rigorous 12-week course that required you to be on base at all times. I didn't get to see Catherine and that was hard on us; especially knowing she was just across the river.

OCS was really a pain in the ass, but I got to meet some great people.

Norman Matzke and I became instant friends. He was a real hard ass; a "soldier's soldier." I liked him instantly. He was the only person there who was older than me. He was a Sergeant First Class and I had recently been promoted to Staff Sergeant. Norman and I would remain friends for a very long time. We both possessed the same dogged mentality.

Sacrifices by family members are unheralded. Catherine had to live, essentially alone, for 12 weeks in a new town. The pressures upon her were immense, many of which I would learn about later in life, but she would never share.

On January 24, 1990, I received my Honorable discharge as an enlisted soldier. On January 25th I graduated from OCS and was commissioned as a 2nd Lieutenant in the United States Army and Catherine pinned on my gold bar. We had done it!

I immediately went to go and join the Pathfinder School, but I had missed the process. I managed to get into the class though because of my prior military service record and I loved it!

Pathfinder School is a three-week course designed to teach soldiers everything there is to know about navigation, creating airborne and helicopter drop zones in the darkness or day, while also showing soldiers how to call in support and understand air traffic control. Here you're taught about being the "first in and last out."

I distinguished myself quickly as an adept and reliable navigator. Everyone called me "Lieutenant Tenacious." The course required that you compete a nearly impossible set of land navigation skills at night with only a paper map and lensatic compass. Instructors would give you a map and a compass and tell you to go and find 6 4"x 4" green painted posts in the middle of the woods, each 4 kilometers away from each other, at night, alone, You had to write down the number of the post on your card.

Only a compass, no electronics.

You needed to locate 6 posts and then combine the numbers taken off those 6 posts which identified the grid location on your map for the final 7th post, which prevented cheating of all types and kinds. This had to be completed in 4 hours.

I couldn't get enough of this kind of stuff. I had always been a problem solver, and this played right into my analytical way of thinking and approaching things.

By the time it was over, I was headed to Armor School in Ft. Knox, Kentucky.

I was excited that Norman Matzke was there with me. He was the epitome of an Armor Officer. Armor Officers are a different breed of soldiers altogether. It takes a certain mentality to climb into a tank and patrol foreign landscapes knowing good and well that your options for escape are limited if shit got hairy.

Being selected for Armor Officer School is a big deal and my classmates made sure to thump their chests and let everybody know it.

"We are fucking TANKERS!"

This bravado was everywhere, and I liked it, though, I found Armor School to be somewhat mundane and repetitive and way too structured for my liking at the time.

I had been in Honduras working in Intelligence and riding around the jungle. We had rules, but we sometimes made them up as we went and if you screwed up, you fixed it and kept on moving. We were cavalier in our movements, but not reckless. Being a Tanker requires the utmost attention and adherence to the rules. It was mechanical not just because you were dealing with a large piece of explosive and destructive equipment, but just in the approach of the cooperation required between Cavalry (The Scouts out in front) and the Armor (The Massive A1 Abrams Main Battle Tanks the Germans nicknamed 'silent death'). A symphony of movement and communication would be required in battle and that made mechanics matter.

Still, I got used to it quickly and learned to navigate my way.

Catherine and I loved Kentucky, and we decided to move to Radcliffe. I drove up to our new apartment in our U-haul. Catherine ran inside to avoid the torrential rain that was coming down. That shit was worse than Tampa any day. We couldn't wait though. Money was scarce and we were on a time schedule. I hopped out, soaking wet and started unloading boxes to Catherine.

"I'm Tommy Ray Hensley. I live next door."

I looked up to see a man wearing a GTE sales jacket and tie. The rain didn't bother him.

"Adam Rousselle."

We shook hands.

"Can I help you?"

Before I could answer Tommy Ray grabbed a box and walked it over beneath the covered entrance to the apartment building.

He smiled at me when he came back for another box. Water dripping from his forehead. I smiled back and from that moment on Tommy Ray and I became life-long friends.

Days later I met Michael Reiber, and we became great friends with he and his wife Tiffany. Michael would later become a part of my company and be an unknowing participant in our dealings with the Mahogany Mafia.

Life was good in Kentucky.

Meanwhile, 6500 miles away, tensions were growing between Saddam Hussein-led Iraq and Egypt and other U.S. allies, including Kuwait in the Persian Gulf Region. Iraq had operated with a license of autonomy for years and the Arab world finally began to speak out. The U.S. issued stern warnings to Iraq for its human-rights violations.

Hussein made threats about the usage of chemical weapons against Israel and the U.S. was forced to begin economic sanctions against Iraq. This infuriated Hussein and war appeared imminent.

In Kentucky, this was literally a world away.

Soldiers talked about it some, but no one truly understood what was going on. Much of the attention in the U.S. had been on the fall of the communist nations in Eastern Europe.

"You think we're going to fucking fight?"

Matze asked me that question one day, and I played it straight.

"Heck if I know."

I didn't know and even if I did, I'm not sure what would have changed in my way of thinking. I was a dynamic soldier. I had been trained by some of the best military personnel in the world. If there was going to be a war, then screw it I'm all in, I was going to fight.

Besides, I knew Matzke would be ready and willing too, so there was no other way to look at it, at all.

As soon as I completed Basic Armor School, we had to choose or request unit assignments. I had no idea what were good or bad assignments nor what armor units there were and what differentiated them. The only thing I could understand is where they were: Korea, Alaska, Texas, etc. While everyone was putting their requests into Armor Branch, I was told, and didn't ask to butt I got picked to be a Scout Platoon Leader in the prestigious 7th Calvary! This was the same Regiment that Colonel George Custer had once led during his famous last stand at The Battle of Little Big Horn.

Being with the 7th Cav was a tremendous honor and reward to be a part of something so historical and legendary. I figured that my past experiences and having the Armor Basic School Training under my belt led to me being selected for such an esteemed post. The 7th Calvary Regiment would utilize my reconnaissance, dynamic thinking and navigation expertise. This was where I was supposed to be. I enrolled in the 4-week Scout Platoon Leader Course., right at Ft. Knox.

Catherine had really settled into Kentucky We had a strong friend network that included Tommy Ray and Denease Hensley. This was important to me because I knew that at some point I would be called upon for duty and I wanted to make sure that Catherine had a strong support system.

Being the spouse of someone in the military was a difficult and sometimes lonely life. When I had to go away for my training courses or classes, I wondered how Catherine was doing. We talked daily, but still, I was always concerned about her. She would always put my fears at ease by telling me she was okay, and I could tell through the phone that she was smiling. Those were

always relatively short trips anyway, and I always came back home the same as I had left.

I guess in a way this was preparation for the life that was to come for the two of us. We were young and still trying to figure ourselves out but there was something big on the horizon. Something that neither of us would've ever thought possible.

It was a warm and muggy evening in early August. I was stretched out across the couch and Catherine was in the kitchen. She came and sat beside me, and the news was on. We watched as Iraqi troops fired long-range weapons and tanks rolled onto Kuwaiti lands. Saddam Hussein had had enough.

"Do you see that!"

Both Catherine and I moved forward to get a better look at the screen. President George H.W. Bush stood at the podium and looked gravely out into the faces of millions of Americans.

"Just 2 hours ago, allied air forces began an attack on military targets in Iraq and Kuwait. These attacks continue as I speak. Ground forces are not engaged."

I don't think I truly understood what was about to happen. It was August 2nd, 1990, and I would never be the same person ever again.

The phone began to ring.

Chapter 16. Battling Apathy

The Army gave me 72 hours to report to Ft. Hood and the 7th Calvary. My mind instantly ran to Catherine.

"When would the Army come and get her?" "Where would she stay while I'm gone?"

"If she stayed here was there some sort of system for the wives to all get together?"

 I quickly learned that none of this existed.   I was headed to fight in a war in the Persian Gulf and my dear Catherine was going to be alone.

We were totally unprepared for this type of deployment. And I felt like it was my fault.

The next few days were filled with angst and worry for me. Our apartment complex was just off the military base. It was a mix of people of all sorts of races, and economic backgrounds, but they were generally nice people. Our neighbors' kids played in front of our door all the time and I loved them.

I had met their dad one day when he was outside firing his gun up in the air.

I was a military man, so guns didn't scare me, but hearing a pistol in a residential complex was unnerving. Catherine and I had been sitting together watching television, when we heard the gunshots. I peered out the window and there he was in broad daylight, holding his pistol pointed up in the air. He was a medium-build African American Man and I decided to go outside and check on him.

"Hey what's going on?" I smiled as I approached him.

I startled the shit out of him as I approached and I'm sure he was thinking, "Who the fuck is this white guy coming up to me and I'm holding a gun."

"How are you doing?" I asked him another question just to let him know I came in peace.

I think this disarmed him some and he tucked his pistol away.

"You must be new around here."

"I am. I'm Adam."

I extended my hand, and we shook hands. He didn't really bother to introduce himself to me He seemed to look around to see if someone was watching this crazy guy named Adam.

"You military?"

"Yes. 2nd Lieutenant in the Army."

"Officer."

"Yes sir."

This really got this attention, and I could see him relax.

"It's really important that we watch out for each other. Watch out for the kids around here…"

"Amen." He nodded, agreeing.

"… We're neighbors and neighbors need to make sure we're safe."

"Yes sir."

He looked at me again. Studied me, as if assessing whether I was bullshitting him.

I absolutely wasn't.

We talked for about 15 minutes about the need for all of us to have a safe community and for the children to have a place to play.

"I will make sure to keep an eye out for your kids."

It was true. I would absolutely do that, whether he was holding a gun or not.

"Maybe you shouldn't shoot the gun in the air anymore. It can scare some folks and you never know someone might get hurt."

"You're probably right."

He nodded and smiled. I smiled back and like that, we had mutual respect and perhaps were friends of sorts.

As I left for Ft. Hood, I thought about him, protecting our part of the apartment complex. It gave me a little comfort to know that at least Catherine would be safe, but my heart ached to think that I wouldn't be there with her.

I thought again about it and later asked Catherine what she wanted to do. As if I could make decisions like that myself anyway, and she most certainly wanted to come to Fort Hood with me.

Ft. Hood was this massive base located in Texas about halfway between Waco and Austin. It was the most populous base in the world and because of its land mass, was ideal for tank training and exercises. I was a 2nd Lieutenant fresh out of some of the best training that the Army could offer, and I had high expectations of myself and my unit.

I was sadly disappointed by what I encountered when I arrived, and I was unprepared and untrained in how to best manage the circumstances I found myself in.

My unit was not fit to fight. We had 60 days to prepare before we would be deployed to the Persian Gulf. I was sickened at the laziness and overall lack of gumption that I saw from my Sergeants. I came in prepared to do my part and take charge as necessary because that was my job, and I took my job very seriously.

The Army is about performing the same tasks over and over until you perfect them. There's beauty in this sort of preparation. Early on, I realized how easy it was to do these things more efficiently and correctly. This was my expectation of all my soldiers because this is what I expected of myself.

I believed in showing up on time and keeping your uniform starched and pressed. Soldiers chose whether to know the 15 things necessary to be in the infantry. I wholeheartedly believed that you had to have pride in being on time and doing the right thing. We were volunteers after all.

I loved being in the Army. I loved being a soldier. I was very proud of this, and if you didn't love being in the Army and being a soldier then you needed to get the fuck out and move on. That was just who I was.

These Sergeants didn't seem like they enjoyed being in the Army at all.

When they met me, they all had that same look, "Who the fuck does this guy think he is?"

If I asked them a question, they'd grunt with a half-hearted answer. If I came into the room, where they'd be sitting and smoking cigarettes, they'd stop talking and look at me. They wanted to establish that though I was the Officer there, this was their show.

I was still young. These guys were older, but our lives were headed in completely different directions in terms of our military careers. They were Non-Commissioned Officers or NCOs. I was a Commissioned Officer, who had once been an NCO. They didn't know that. They never cared about becoming officers, nor would they have likely qualified. But worst of all, they didn't give a shit.

"Where the fuck you come from?" an E-5 Buck Sergeant said to me, in front of the E-6s and my Platoon Sergeant E-7. I counted to 8 in my mind.

"Excuse me?"

I had to look the Sergeant in the eye to really see if he was asking me this fucking question.

"Who the fuck are you?"

I walked out of the room to collect my thoughts, and concluded I needed to learn more about these Sergeants. Fuck that, this is not the Army I belong to.

I was two days in, and I knew I needed to get this shit straight, ASAP.

I poured over information about my unit which included the Sergeants and many Enlisted Soldiers, including 8 Privates. What I found out was that of those 8 Privates 6 of them were at least 18 months behind on their promotions. I was crushed. This was pure bullshit on the part of those NCOs.

As I jotted down my notes and pulled more files, I could feel my blood boiling. I wasn't a big guy in stature, but at that point, I wanted to kick the shit out of all of them. I loathed them mutherfuckers for their incompetency and unwillingness to look after those who served under their jackboots.

These guys had a self-centered vision of the universe. They had been in the Army for so long that they felt like there was no need to care about others because there wasn't a need. There was no performance needed from their subordinates.

The vehicle maintenance was in shambles and because they drank with the motor pool chief, their maintenance checks never surfaced to cause them a problem. It was one big playground and they owned it. 7Th Cav Scout Platoon Sergeants were at the top of the heap, and they wrote their own rules. In fact, they griped and smudged over the Army's leadership ethic value so often the young soldiers who might have had a bright light burning when they arrived were all but snuffed out by now.

They knew that as Sergeants, if they wanted to get promoted, they would go before a Board. The Board would look at their time served, ask them competency questions, and then grant their promotion, as long as they had not already committed a crime or anything generally stupid. The other younger Enlisted guys were on automatic promotions. You'd go from Enlisted Private E1 to Private E2 automatically in 6 months as long as your NCO put in the papers saying you had not yet had an Article 15, then Private First Class in the next six months under the same process, and then Specialist by 24 months; automatically. This was simply based on how long you had served and evidence from your leadership that you were still alive and not a felon.

All these Sergeants had to do was submit the paperwork and get these guys their promotions and money! Many had wives and multiple children. Pvt. Willis was one of those. Lots of kids, a young wife, living below the poverty line, as most soldiers do, and the money would matter a lot to those kids.

I stormed into the office the next day and slammed my folders on the table. Those Sergeants turned to me with a "Here's this fucker again" expression.

I exploded.

"On your feet!" I locked their heels and left the door open. Our Platoon leadership room was in the barracks, and I was sending a Les Thatcher message:

"How could you not fucking promote these guys?"

I read their 201 files one at a time; dates enlisted, date of rank, Military Operational Specialty (MOS), and the lack of any negative reports. Indeed, no reports of any kind and in this case that means promotion as required. PFC Michael Cagle on Track A34, PFC Stan Cebula on Track A36, PFC Michael Condron on Track A36, PFC Tim Frank and PFC Anthony Kuykendall on Track A35, PFC Tommie Mitchell on Track A35, PFC Phillips, and PFC James Willis on my Track A31.

SSG Meyers attempted to sit down, and I wasn't having any of it.

"Lock your fucking heels Staff Sergeant! I'm not done! I looked him square in his eyes behind his dirty, eyeglasses; his uniform looked like wrinkled shit.

"Of those 8 privates, 5 are yours.'

Spit flew from my mouth, and I slammed my own 201 file on the desk.

I read every school every report and every training and every NCOER (Sergeant Evaluation Report of my performance) and OER (Officer Evaluation) about their new Lieutenant and that just 6 months ago I was a Staff Sergeant in the Regular Fucking Combat Arms Army. I knew the rules, their rules, our rules. The rules.

"You're a fucking disgrace to the Army and to your mutherfucking selves!"

They couldn't say shit because they knew that I was right.

"You in my opinion are a lump. You are a lazy, good-for-nothing, lump. And until you straighten out, there is not a performance review that you are going to get from me that doesn't start with the words, "Flat out Lump." No reviewer will misunderstand what I'm trying to say. It won't matter what it says in the title Sergeant, or that the reviewer reading it is your buddy and you all had been hanging out and loving on each other. When an officer says that you're a lump, you go to the bottom of the line behind every other soldier. A Sergeant at your rank and your age needs two positive reviews. I give a review that is negative and there will be 5000 other soldiers in front of you that will be promoted. That's how long of a wait it will be. "

I was on a roll and the room was dead silent.

"My review means more wages for your family and your kids. My review means that you get promoted. If I give you a bad review, you're an E5 today if you make it to E-6 it'll be over my dead body and if you're an E6, E7 it'll be when I'm 50!"

I treated them just like they treated Private Willis.

They hated me for that, but it was true I had to be an asshole because there had been too much bullshit going on for far too long. I made a choice, and it was scary to go-it-alone so to speak. However, I didn't know how much of a choice I had.

Three weeks to Saudi Arabia, as an officer I had to set my parameters and I did. But they weren't going to change easily.

I learned in that moment how to Lead by Awareness. Like leading by example, however, finding better data, more relevant information on the enemy's new weapons, better maps, back-up methods of navigation, finding out more about my soldier's lives than they were willing to share was powerfully important. I could see my soldier's growing confidence that I had sufficient information to make decisions and that in turn gave me the Courage I needed to lead in hard times.

I immediately filed the paperwork to promote all those Privates and got their back pay to date processed and in their pockets. Every fucking penny. I gave them a pinning ceremony and had those Sergeants fully participate.

"Thank you, Sir."

I could see the gratitude on Private Willis' face. He was a young soldier (younger than me) who was eager to learn and perhaps make something better of his life. The problem was he had a stigma attached to him that I knew about. His family had damn near been starving and he and his wife had 3 children. They were poor. The weight of all of that got the best of him. Those mutherfucking bastards had not realized what their role was in Willis' life and that pissed me off.

"You've earned it private We should apologize to you. Remember today Willis when you become a leader."

I was happy to help him, and I knew that at the least the back pay and the salary increase that came with his promotion would help alleviate some of the financial strain he had to be feeling. His salary doubled! My heart hurt for Willis.

I made sure he was assigned to my Track. "Track" is the term we used for our Bradley Fighting Vehicles. We never said, "My Tank," we'd always said, "My Track," "Your Track," or whatever.

Willis was then assigned to Sergeant Dial; Sergeant Michael Dial who was my Gunner. Dial was the Army's reigning Audie Murphy Soldier of the Year award recipient, which meant he was recognized as an outstanding NCO who went above and beyond in their leadership roles, embodying the legacy of Audie Murphy, one of the most decorated U.S. Army combat soldiers of WWII. That was a big deal.

Sergeant Dial was young, but he was extremely qualified. I felt good having Private Willis on our track because I knew that I could help him, and he was smart I knew it. He was articulate and kind. I wanted to keep an eye on him and make sure he was safe.

Willis did what I asked without hesitation. He supported what I asked without question. If I gave him something to study, he would study it until he knew it, then he would study it again and again.. We trained endlessly in that Texas desert and Private Willis, and I would often have conversations in our track.

"Did you learn your flashcards?"

"Yes sir."

"It is going to be crucial Willis that when you look in your reticle you can distinguish which tank is an enemy and who is a friendly."

"Yes Sir. I won't let you down."

No one had ever taught this kid how to make breakfast cereal. He soaked it all up and through all of those hours of training, I gained confidence in him. He quickly understood the maps and navigation. Willis needed to understand the difference between a BMP and a BMP 2 at 3000 meters with thermal sights. If he ever sat in the Turret next to me or Dial, we would need his eyes to understand what we would be looking at and ensure all of our safety. Willis was going to be a gunner someday.

He needed to be confident.

One night when it was deathly quiet and still, we sat there beneath that Texas sky and I told Willis the reality of his job.

"There's gonna come a time when you gotta pull your shift. You're gonna have to be able to look through that thing and tell me which is which. It doesn't matter if I or anyone else likes what you're telling us or not, you gotta have the courage to say it. Even if everyone else is wrong. If you don't say something, someone's gonna die."

One of the things Les Thatcher taught me as a Commissioned Officer and as a NCO is that "if you want to understand soldiers, and who they truly are, "You gotta go and visit them."

This means you go to where they are when they're working and you say "Hi, how are you doing?"

It's really simple, but no one does it. Most officers really don't know what's going on in their soldiers' lives because they don't know what's going on with their NCOs. If they don't check in on their NCOs, the NCOs don't check in on the soldiers and soon the gap emerges. Lead by example, it's really simple, but doing so is nearly a binary condition to effective leadership on post and it's absolutely necessary in combat.

General Colin Powell wrote in his forward for the 1986 Army Noncommissioned Officer Guide:

"If you are going to achieve excellence in big things, you develop the habit in little matters. Excellence is not an exception; it is a prevailing attitude, studying and understanding every detail will maximize your options when time is against you..."

The NCOs are the officer's link to their soldiers. The way the soldiers acted and worked told you what kind of job the NCOs were doing.

I already knew what my NCOs were like, and I also knew needed to check on my soldiers myself.

In the Army, if you ever wanted bullets, mines, grenades, or explosives, you had to go to a special place with a ream of paper and 50 signatures to see the guy who could help you. You weren't going to just walk in and take any ammunition you wanted at your leisure. There was a serious process to touch ammo. But at that time at Ft. Hood, because we were on our way to war, all the rules were nearly off.

You went and got all of the ammo you needed, and you would then bring it back to your motor pool, which is the Army's way of saying large parking lot for anything that moves, and stow it in your tracks. This place had a chain link fence that wouldn't keep a bear out of it. But now it had all of the United States Army's tanks with live rounds, long-range anti-tank missiles, Live 25 MM rounds, M-203 frag grenades, Claymores, rocket launchers; you name it, anything you wanted sitting out in the open behind this bullshit barrier.

There were missiles in there. Tanks/ Bradleys, whatever. Hell, if you could drive the track, you could probably sneak in there and drive the damn thing home. Hell-of-a-time trying to stop a Bradley on a highway with a police cruiser.

The bottom line is that meant that someone had to guard this shit 24 hours a day 7 days a week.

I went to the Motor pool one night because our platoon was on Motor pool guard duty. At Motor pool guard duty, you are weaponized. You have a gun, and it has bullets in it. I know that we always had rounds in Honduras, but at Ft. Hood, they didn't dole out ammo haphazardly. Problem was these guards had no real training on what bad guys looked like who might want to take our shit.

This is the real deal. You were protecting 50 fully combat loaded tanks and 150 Bradleys in the middle of Texas with no one around and a chain link fence and a class 3 padlock that a crowbar could break.

Tanks and Bradleys don't have ignition keys. If you know how to operate them, you can just drive the mutherfucker away. So, it was imperative that the guards have their shit together.

When I got over there after 1 a.m., one of my newly minted Private First Class troopers was laying down on his back, using his radio as a pillow with his loaded M-16 2 yards away leaning up against the fence. He was on SSG Meyers track. He was the only soldier on guard duty and there was supposed to be four of them!

My platoon, no guards, loaded tanks, loaded Bradleys.

I went bat shit nuclear crazy and for the first time in my life, and this is against the law: I kicked that soldier in the ribs so hard, I'm surprised I didn't ruin his spleen.

He woke up in a panic. I emptied his weapon and ripped his rank off his lapel and yelled,

"You are hereby demoted. Mutherfucker!"

I got on the radio and called in the Desk Sergeant. I got the whole NCO leadership down at Motor Pool. "Who is his squad Sergeant? You got 30 seconds, and you better have a great fucking excuse or you're gone in the next 30 minutes!"

We were not combat ready.

I waited, pissed off, for someone to show. When the Sergeants began to arrive, I lit into them.

"We are to going to combat, sleeping with loaded weapons. Near the enemy. What if someone broke in? or simply stole the loaded M-16? Jesus! If the Colonel were here the Division Commander might chew his ass. The entirety of the 7th Cav chain of command would be a laughingstock for such incompetence, your incompetence. The ONLY FUCKING reason I knew this is that I ALWAYS check on my troops. You stupid mutherfuckers!"

It was now 3 o'clock in the mutherfucking morning. I had started calling people at 1:45 and I was pissed that it took them that long to get down there. I had the whole platoon on guard duty.

Staff Sergeant Meyers tried to say something.

"You lost the ability to speak to me about anything as an NCO for not promoting those privates. Tonight, Your soldier fell asleep with a loaded weapon on guard duty. You had better turn the fuck around and think about your next words wisely."

That was it. My soldiers could have said a lot of things about me, but one thing was for certain, I cared, and I knew that rules and words mattered and that I following the rules. And I didn't care at all about going along and getting along.

When I left them, they gave me a plaque about self-less service.

Eventually our unit began to round into shape. We started to look the part of a combat-ready Cavalry Troop.

It wasn't long before our orders came in for us to leave. We were going to war.

Chapter 17. Shut Up and Show Up

We landed in Saudi Arabia at the Port of Dammam in late October of 1990. We spent two weeks there before moving north to King Khalid. After a brief stay there we moved north to our Screen Line. A screen line refers to how a Cavalry unit would deploy side-by-side with as much as 2000 meters between vehicles to expand our ability to cover terrain while staying in contact with each other, allowing just short of the berm that separated Saudi Arabia from warring Iraq.

Our entire operation was part of a joint international coalition of countries, many from the Arab Region who opposed Hussein and his tactics. I was tied into the Egyptian Armor unit as the U.S. officer in charge of communicating with our Egyptian counterparts and coordinating our military activities.

Once again, I was the guy with the foreign connections, only this time, unlike in Honduras, these guys were in tanks, and they were right next to me. I relied on them for protection and vice versa.

This made me nervous as hell.

The Egyptians used our old T-60 U.S. Army tanks that had rust all over them and they looked like they needed to be junked immediately. It was scary. Their uniforms were disheveled, and these guys didn't know what the hell they were doing. In combat, you had to be ready, or the enemy was going to shoot and kill you. These guys weren't ready, and I knew that if we were in combat and they got killed; which seemed highly likely, then my ass would be out there flapping in the wind with a bullseye directly on me. I didn't want to die, and I didn't want to lose any of my men.

This was mental torture for someone like me. I was all about the details and knowing everything about my Track and my missions. These guys didn't know if the barrel of their gun was warped! I didn't know who they were. I didn't know what kind of mental or physical condition they were in. I didn't know what kind of condition their rusty tank was in.

Had they ever tested the mutherfucker to see if it actually worked right?

But despite my reservations, I had a job to do. It was an honor to be chosen as the American officer tasked with communicating with our Egyptian allies, I knew that I got the job because someone up the chain of command thought I was the best Armor Lieutenant that we had. The Colonel had decided that I would be the guy on the ground to talk to their guy on the ground. I was the one that had to tie our two armies together. That was a big deal during wartime activities.

I didn't really recognize the value of this appointment at the time, but it would dawn on me many years later. I was so focused on doing what I was supposed to do that I didn't realize that I might actually be doing a great job at it.

But I had more than just my Egyptian counterparts to worry about.

Saddam Hussein had made it clear to the world that he was prepared to use chemical weapons against anyone he perceived as a threat to his regime which included his own people and all allied coalition forces. This meant that U.S. troops needed to be able to defend ourselves against chemical warfare.

As a result, each soldier in our platoon was prescribed, Pyridostigmine-bromide (PB) which was an anti-nerve agent pill used as a pretreatment to protect soldiers from death in the event Hussein attacked us with the nerve agent, Sarin. In other words, we were being given doses of the nerve weapon to build up immunity to it should we be exposed to it. We were instructed to take the pills and told that we may experience some sickness.

I got ill within 7 hours as did several other soldiers. My abdominal pain was intense and unrelenting. At 3 a.m. I was dry-heaving and in unstoppable pain. I knew this wasn't normal. I was in the best shape of my life and had been on a sterile diet of MRE packets (meals ready to eat) 3 meals per day for months. This wasn't supposed to happen to me.

With some reservations, I made my way to the Flight Surgeon. My experience with Sanocki had made me leery about seeing field medics unless it was absolute dire straits. This was one of those times. The pain was simply unbearable. I had to go.

"How long have you been feeling like this Lieutenant?
"It's been a few hours now Doc." "Fuck!"

I doubled over.

He poked and prodded around my abdominal area. He could see that I was in immense pain, so he gave me two painkiller shots that quickly knocked me out until the Major showed up.

When the Major came in around 6:30 a.m. he was in disbelief. He went batshit crazy when he saw me.

"Oh my god. What the fuck is wrong with him?"

I had a fever of 104, with violent chills, and he knew I was really sick. I was a combat line officer. We were in the midst of war. An officer like me would never go to see the Flight Surgeon. There were people who needed me to be in place; folks that relied on me.

"It's his appendix."

The Major turned to the Flight Surgeon.

"It has to be taken out."

On December 8, 1990, I had my appendix removed in a tent in the desert in Saudi Arabia. Had I not had the surgery then I would have died when my appendix burst. I had managed to escape death once again, but it wouldn't be the last time.

I was medevaced to the U.S. Naval Hospital in Bahrain for recovery. I wanted to get out of there as quickly as possible and get back to my Unit.

"With your recovery, you can return stateside."

The Naval Doctor assumed that I would want to go.

"Can't do that Doc. Gotta get back to my guys as soon as possible."

"Understood Lieutenant."

"But I do need to get out here for a break. Tired of laying in this bed."

The Doc nodded. He scribbled some notes on my chart and in my file and left.

An hour later a nurse came by.

"Do you want to go to the Officer's Club?"

I looked at her in disbelief.

"The O Club? You have an O Club?" I popped up. Already getting to my feet.

"Yep, it's not much but it's a few blocks from here are you can walk there, but you'll need to take off the hospital wrist bracelet Lieutenant" She took scissors and cut it off and smiled kindly. She handed me a replacement bracelet and I left.

Trouble was, while your soldiers are in a forward position in a Muslim country, you are not allowed to drink alcohol. If you do, you will lose your Officer's Commission. It's a serious offense, but I figured there was no way I would get in trouble on the Island nation of Bahrain.

I walked into the O club ready for my drink. The nurse was right. There was not much to the place. It was an old contractor's trailer that had 6 seats which was 1 more than the choices of alcohol. An old VHS played reruns of an NFL game.

I ordered Makers Mark and Ginger Ale. The effervescent bubbles made me sneeze and I smiled wide.

Suddenly, the screen entrance door slammed, as it had when I entered. It sounded like grandma's screen porch door. A grey-haired man sat down next to me; our shirtsleeves touched since the fixed seats were so close.

I was in deep shit.

The ONLY people allowed in an O club are officers. AND the only grey-haired officers, are usually Generals or Admirals, and they knew the rules for officers.

But it got worse. The door slammed again!

Another grey-haired officer sat down now to my left.

I was fucked.

Fortunately, (maybe) I could tell they weren't Generals. But they could easily destroy my career the same. The Vice Admiral to my right, leaned forward and looked at my Army BDU's and said.

"Lieutenant: "You're a long way from the front, aren't you?"

"Yes Sir. That I am."

Then the Rear Admiral to my left asked, "You broke out the hospital?"

"Yes Sir, I did."

I pulled out my un-used replacement bracelet as evidence.

The Vice Admiral chuckled. "What are you drinking Lieutenant?"

I reached out to my glass of Kentucky heaven with both hands as if I were warming my hands with a hot cup of coffee. This was going to be a very bad day.

"Makers and Ginger Sir."

I knew I was caught. The rear Admiral shook his head. Damn.

The Rear Admiral looked at me. Then motioned to the bartender.

"Give us what he's having. And give him another one."

I almost fell out of my seat.

The two Admirals laughed.

The Vice Admiral patted me on my back.

"Son, what happens in the O Club, stays in the O club."

I got back to my hospital cot in one piece and reinvigorated with the belief that honor existed within the Officer Corps.

A few days later the Doctor removed my staples and I checked myself out of the hospital with one thing on my mind: Get back to the 7th Calvary.

I was used to making my way around places I didn't know and having to use my ingenuity to get what I wanted. On my way out of Bahrain I "grabbed" 42 immersion heaters from a "friendly installation" (it was December after all and maybe I didn't necessarily fill out all the proper paperwork, but who cared, we were in a war), and I hijacked a 2 ½ ton truck and drove my way. I didn't have time to wait for the next caravan to take me back to my Unit.

When I got there everyone was shocked as shit to see me. You would've thought I had come back from the dead or something. The general assumption was that I was going to take that opportunity to return home. Some of the soldiers had already split up my belongings. I understood that. It wasn't like they were about to send my shit home to me on a UPS truck. No one was going to lug around the Lieutenant's supplies so he could have a keepsake when he returned.

I loved my wife dearly and I desperately wanted to see Catherine and hold her again, but I knew that she would want me to be in the Persian Gulf to finish what I started, and I wholeheartedly wanted to be there.

I rounded up my shit and got back into my groove pretty quickly. There was no time for pleasantries and relaxation. The enemy was just over the border. We could see them in plain sight sometimes. We could see the flashes at night of their heavy artillery in the distance. It was unnerving at times to know that at any moment we could be attacked.

A few miles in front of our screen line the Army had these massive speakers set up that blared all kinds of psychological warfare messages. Sometimes simple words and sometimes unrelenting horrific screaming. The Psyops guys blasted that shit 24/7 to screw with Saddam's troops.

We didn't have daily fights with Iraqi tanks. That wasn't the nature of our mission at the time. We were trained to "Find-and-Fix" the enemy's locations. That's what the Cavalry does. We are scouts, like Custer was. The captain would regularly update what was known about the enemy's locations and post these figures on a board or plotter that would show us on a map where they were. If we deployed this would give us a head start at ensuring, we could track (find) them down and make sure we knew their every move (fix).

If they moved one mile backward, we moved one mile forward. If they went to the right, we went to the right. We were relentless in our efforts, and preparations so that when the actual battle came the tanks would move forward and wipe them out.

I had been positioned as the center platoon leader of Alpha Troop 1st Squadron 7th Cavalry. That meant I was in the middle and the most forward officer in the 1st Calvary division. My job was to ensure that no enemy would pass through our screen line, and otherwise to "find–and fix" the enemy if we ever saw him. We would look for the enemy with our thermal night sights. If we ever saw movement, we would report back to HQ on his every move. We never let them out of our sight.

But this was very taxing on our soldiers, because to find the enemy we had to use our thermos night sight at night and that meant you must stay awake all night. Someone always had to be up, so we were sleep-deprived and worn out.

I knew that something needed to change, or we simply weren't going to be able to do our jobs. I asked our Squadron Commander Colonel Sharp to assign to my Platoon a counter-battery radar truck which wasn't doing anything in the rear but collecting dust. I reasoned that if we could use radar at night, my soldiers could get some rest. The radar would go off if we found the enemy.

The Colonel agreed and it was done.

The Pipsi-50 could "see" 25 miles away, while our thermal night sights could only see about 4,000 meters. This allowed us greater confidence to let our troopers get some sleep and also be capable of responding to any threats quickly. So, this night, the guy's got a radio and he says, "There's a movement out there."

I radioed back.

"Are you sure?"

"Yes. There's a bad guy out there."

In truth, we didn't know if he was a good or bad guy. I knew from my experience with the 7th Special Forces Group that Green Berets often went on stealth missions at night, and no one knew about it without the proper clearance. They'd do shit like cut supply lines or sever communication lines just to screw with the enemy on their own turf, right in their own backyards. I had an appreciation for this, so I didn't panic.

We positioned our guns and got ready. But you don't just shoot at somebody in the dark without the "ok" from Command.

I called it in.

"I think I got a bad guy up ahead."

I gave the coordinates to Lt. Bagdasarian, our executive officer, and he called them into the Colonel. The Colonel then picked up the radio and called someone else like a Les Thatcher type guy who oversaw the Special Ops.

We were getting antsy. These guys were coming towards us and that's never good.

"Any word on these guys?"

The Colonel listened and waited for a reply from the Special Ops and Intelligence guys. We all held our breaths waiting. Was this the moment we had prepared for? You could've heard a pin drop it was so quiet.

And then... nothing.

We got no fucking response. Shit happened like that sometimes. A message might get mixed up in the communication and ultimately no one ever got the right answer. I made the call to simply monitor and report.

Hours went by. I kept tracking the enemy. It was tense and frightening. You simply didn't know what would happen.

Being in a tank in the middle of the night can be a lonely feeling. You're limited by what you can and cannot see and do. You're vulnerable in a sense to enemy fire when they get within a certain distance and at this point these guys had long passed that point. We were there alone. Or at least that's what I initially thought.

Our Troop had a platoon of Mortars. These guys were Infantry soldiers. Not Calvary Soldiers in tanks like us. They were essentially foot soldiers with heavy mortars. They had a culture that I loved, but they didn't like the Calvary guys. There was sort of this belief that the Calvary guys weren't as tough since they rode around in Bradleys. They wouldn't hang with the Calvary guys. They wouldn't drink with them. There was definitely a separation between the two groups.

The Infantry guys lived by the code of "Shut up and Show Up." They believed that you do what you gotta do for your guys without asking questions.

That was exactly what I believed.

One of the guys got on my radio frequency.

"Yo Lieutenant. It's Rankin."

My shoulders relaxed when I heard his voice.

"Good to hear from you Sergeant"

"Hey, Lieutenant, we moved our platoon up just behind you. Say the word and we'll blow those bastards to hell."

The hairs stood up on my arms. That was what it meant to be a member of the United States Army and be a part of the brotherhood.

Nobody gave him the order to do that. I didn't ask him. And he didn't ask the Captain, but Sergeant Rankin heard my calls on the radio, reporting what could be bad news, and he got where he needed to be.

He showed up and shut up.

He heard that we were tracking someone, and they were getting close. We couldn't easily fire our weapons because the blast from our 25 MM guns would give up our locations. The mortars on the other hand could fire from 1 km behind us, dug-in and out of sight, and the enemy wouldn't see their tubes fire nor stand a chance. Even if they saw the mortar blast, it would be too late. Shit would have hit the fan and everyone was going to die.

But he also had to put his troops in danger by moving that close to us. Once he set up his mortar tubes, taking that shit down took time. If we had to retreat or get back, he would be left out in the open with little protection. It would then be up to us to try and stand and fight and help him. But that didn't matter to Sergeant Rankin. He was there for us and I truly appreciated him and his men.

By the time he radioed me, he had already dismounted his mortar platoon in a dangerous place, dug-in and dialed his tubes in, because if I was right and this was the enemy, we were going to need them.

The Army never acknowledged those guys for what they did. That selfless act was the epitome of what was asked of our heroes. It was truly a selfless act of service, and it pissed me off that they never received their recognition for it.

That night ended peacefully. The enemy decided to turn around and head back the way they came. All I could think about as I tried to decompress from our angst was how grateful I was for Sgt. Rankin. He "showed up and shut up."

Chapter 18. There's No Music

As a soldier, particularly during wartime, you are in a constant state of mental flux. Anxiety and fear are compounded by the mere fact that you "just don't know when IT will happen." That "IT" is the firefight. As a soldier, you have committed to laying down your life for your country, family and friends, and fellow servicemen and women. You aren't necessarily afraid to die; no, the real fear comes in never knowing when combat will happen and how to control IT to your advantage. Like Napolean stated, choosing where and how to engage is as critical as engaging at all. Platoon leaders do not get to choose where to fight, that is the role of the Squadron Commander. However, if you study, anticipate the permutations, and prepare for what is likely to happen, you can enhance how you execute your attack.

This sentiment is heightened for officers, at least the ones that care. You ride the ebbs and flows of your emotional roller coaster, always analyzing, always studying every new bit of data as the permutations of what it means, and you also carry the weight of wanting to protect your men against harm. I didn't want anything to happen to my guys, especially Private Willis, because I felt particularly bad for him. By no fault of his own, his Sergeants had not prepared him before my arrival for what he now faced.

He was eager to learn and as I talked with him and quizzed him about his job, I wondered if he would make it home safely.

I was going to do everything in my power to make sure that he did, as well as the rest of my men.

These feelings are the soldiers' dilemma. Some people block these thoughts out of their minds completely. Others resort to physical conditioning and working out. I chose to prepare. I studied everything, every bit of information that we were given about our daily tasks. I knew where I was always supposed to be with my Track and where everyone else should be. I became obsessed with preparation.

When the time came for the firefight, and if I were to die, I was confident that I would die knowing everything I could to save those around me and it was simply my time.

In late February of 1991, our missions took a dramatic turn. We were now going to engage our enemy.

"Gentlemen we are going to Basra, Iraq. We have orders for a movement to contact."

The Captain looked stoically at all of the officers gathered in the room. My heart rate immediately quickened, and I felt the adrenaline shoot through me. This was it.

"Movement to Contact" was a big deal. It meant that we were going to move until we made contact with the enemy. It was either going to be a fight or they would have to surrender. Once we started moving you won't stop or aren't allowed to stop until you find the enemy.

That order meant the leash had been taken off and the dogs were being set out to hunt. The Calvary was about to do what we had been trained to do.

I took out my notepad and began jotting down notes about everything that he said. The room was crowded and there were 12 of us there. These were Senior Officers above me. But I didn't care. We led the line-of-march and were slotted to set our screen line with my platoon in the center.

If we made contact, I was literally going to be the first of my unit to confront the enemy in our Bradleys!

Pumped with adrenaline, I began to elbow my way up front, a lot of supply and rear guard guys were crammed in listening I needed to hear every word that was being said.

"Excuse me sir. I'm your scout platoon leader."

That's what I told the planning Major. I had no fear. I had to ensure that I heard all of this. They made room for me and I took a knee,

"It's a 3-day journey to get to Basra, Iraq. But we aren't taking the direct route."

I leaned in. My pen was working like crazy.

In an ingenious strategic move, General Schwartzkopf had decided that we were going to go around the enemy that was directly in front of us!

I excitedly squeezed my pen, almost breaking it in half. My teeth clenched.

"Here's what we're going to do..."

This was perhaps one of the most brazen and smart moves since the Trojan Horse was wheeled dead smack in the middle of the city of Troy.

We took giant pieces of plywood and stapled electric bed heaters from Walmart to them. Damn, Walmart bed- heaters went to combat for America! Thanks Walmart! We then placed generators that powered the heaters behind the plywood. But that wasn't all. We then got a giant engine that ran oil out of a manifold to make smoke that looked like the smoke that came off a Bradley and then glued mirrors from Walmart onto pieces of plywood. So if the enemy used binoculars he would see what looked like sight boxes from the top of our Bradleys. If he had thermals to detect heat sources, he would see two hot things that looked like our engines were running.

The enemy would think we were still sitting still when we were actually driving around his flank to ambush his rear!

We set these things up and then backed up 3 kilometers and went all the way around. The enemy was right in front of us, but we had faked the Iraqis out and they had no idea what we were doing.

The 3 day journey was exhausting. We only stopped for fuel.

One day, I drove over an enemy dug-in infantry platoon. Our reconnaissance guys hadn't told us about them, but we literally drove right on top of their asses!

I ordered the platoon to dismount our Bradleys along with my soldiers. Were they still there? Was this the day? I ran into a bunker with my 45, loaded, and the hammer back with my soldiers armed with their M-16's right behind me.

I then led my guys into a bunker, fully expecting that someone would shoot at us as we entered...

... Nothing.

A lit candle sat in the middle of the bunker. A coat hanger held a tin cup that was steaming!

Someone was in the bunker. We pushed into the darkness. Weapons pointed, waiting on any movement. I could feel the breath of my men on my neck. The tension was palpable and choking.

We all knew that at any moment, we would die.

"Nothing!"

Someone yelled out. It was all clear. We had just missed the enemy soldiers by minutes perhaps. Maybe they knew we were coming.

We got back on our tracks and kept moving.

That was a reality check for all the soldiers I had chastised back at Ft. Hood. They realized that this shit was real. We weren't here to play games. If someone had jumped out in that darkness there would have been casualties, perhaps on both sides. It dawned on them that they may not have been prepared. Some had chosen an easy route and didn't adhere to my advice about "constant preparation."

It was too late now. There was no turning back.

February 27, 1991

I hadn't slept for days. Our radar no-longer available to track the enemy, as a lead officer, I had to stay awake. Plus, sleeping while others worked wasn't who I was. I pushed myself mentally and physically until I could tell that it was taking a toll on me. I needed to get some sleep as soon as possible.

I turned to my gunner, Sergeant Dial.

"I'm going to get some sleep, but I need you to wake me up before the Captain gives out his instructions."

"Yes sir Lieutenant."

"6 hours. Once we're close, but before he positions us, wake me up."

"I've got you covered."

"You're in charge."

I trusted Dial. He was a good soldier and I figured he could handle things while I slept.

During our earlier movement to contact, Dial spotted a small smoke stack protruding from the desert floor. Essentially it was a stove pipe through the roof of a concealed bunker and we were about to drive by it.

"Driver STOP!" Dial Ordered.

In an instant the front sprockets locked up and the Bradley stopped dead. I radioed the platoon and they followed suit. We couldn't bypass the enemy observation bunker because that would expose our non-armored rear doors to the enemy. If they had an RPG, it could get real bad. We had to deal with this threat and we were the lead Bradley.

Sgt Dial had an idea, it was risky, but he was brave and resolute in his suggestion. I contemplated his approach and agreed. His plan was to drop a hand grenade down the stove pipe and run.

This meant he had to literally get on top of the enemy bunker and hope they couldn't hear him. By now they had to know we were near.

Sgt. Dial grabbed two M-67 Fragmentation grenades and I gave him my 45 Cal Pistol. The plan was for him to drop one grenade down the stove pipe and keep one in reserve. I would man the 25 MM and over watch his position. If any threats came up on him, well, they wouldn't stand a chance.

Dial stealthily approached the enemy bunker, crawled to the stove pipe, rolled onto his side, and looked at the turret. He couldn't see me, but he knew I could see him close up. He showed me he pulled the pin, let the spoon go and counted to one, dropped the grenade down the pipe, stood up took two steps, and dove into the sand with a belly flop.

BOOM!

Dial stood up, walked back to the Bradley like nothing happened.

Dial was indeed a great soldier. We would be fine under his watch.

I climbed through the turret and into the back of my Bradley and within minutes I was out like a rock.

I had to hear this briefing. The Captain was going to tell us our location, and where to set up. Where we would attack from. When the Captain gave his briefing it was crucial to ask questions.

I always asked questions because I needed to know. "Who is to my left?" Who is over there?" Only the Captain had this information and I needed to have answers to those sorts of questions to fully know what was going on and to feel the most comfortable to lead my men.

Dial never woke me up.

When it came time for the Captain's orders, the Platoon Sergeant told him not to wake me up.

"Let that fucker sleep." I could only imagine what he said.

He didn't want me asking questions and doing my job. Some of those guys simply wanted to get by. They wanted to see how quickly they could get off the radio with the Captain and go from there. He didn't realize that my questions were there to save all our asses from getting killed in the middle of the Iraqi fucking desert!

"He's my boss. He told me to let you sleep." Dial tried to plead his case.

I could've blown a gasket.

"Sergeant. Do you understand what the fuck you just did!"

So instead of telling me what to do the Captain told my Sergeant First Class, a 51-year-old asshole where to position the platoon, but he, in his infinite wisdom, failed to ask any other questions that I needed to know.

"C'mon Sir. I only did what I was told."

"Bullshit. You're MY gunner and I told you to wake me up. And you didn't. I trusted you to do one goddamn thing and you didn't do it!"

I wasn't done with Dial yet though.

"We're in an attack position. I gotta jump us forward and I don't know where everything and everyone is!"

If I could've kicked Dial out of the tank at that moment, I probably would've. But I did the next worse thing.

"Get out of my turret. Willis get in here!"

Without hesitation, Willis stood up to take his position. Dial was incredulous, but he had no recourse. His actions had put us all at risk and I didn't give a shit about his feelings at that moment. The fallout from that mistake would haunt Dial for his entire career and he would never recover.

I was too harsh on Dial, and that day would also haunt me.

My mind raced. "Where are the other Lieutenants..." I asked myself this question out loud, knowing that I was the only one who had even slight knowledge of their locations. Hell, I didn't even know where I was. I hadn't slept long enough and I was still beat. All of the preparation and planning seemed to have been for naught. Now when I needed my information the most, it wasn't there.

I thought I knew where some people were. I started plotting them on my own map in the dark with a flashlight dangling from my mouth. I used our gun to estimate the distance between us and the other Bradleys. I managed to plot where I figured everyone was, but I wasn't completely sure.

We had to pull into a screen line by 9:30 PM and I did my best to line up all the vehicles on my map.

Eventually, I simply fell asleep. My brain just shut down.

I was awakened by a forceful nudge from Private Willis.

"BMP 2 o'clock. 1500... meters! Closing on the Screen line fast Sir!"

I was up and ready. I quickly cleared my head.

Willis had done exactly what I had asked him to do. He was prepared and now he was staring down the barrel of an advancing enemy tank's gun.

I looked through our scope. Willis was right! Time was of the essence. Every meter that that tank moved meant that more and more soldiers could die. I had to make a decision in a hurry. I best guessed which track was closest. It was Sergeant Robert Miller A36.

"6 this is 1, BMP your 12 O'clock 1400 meters, Fire."

"1 this is 6, negative Sir. That's a Bradley. That's Ours!"

What the Fuck? I didn't know what he was saying. Was it a Bradley? Was it an Iraqi tank?

Willis was antsy in the turret now.

I radioed to Staff Sergeant Meyers on A35.

"5 This is 1" BMP your 1 O'clock 1200 Meters Fire."

"1 This is 5, negative, it's a Bradley, It's ours."

"No it's not. It's the fucking enemy!" I yelled to myself.

The moment was chaotic because I knew that Willis knew that that was an enemy BMP (tank) and we needed to stop it before it fired on us or our medics and cooks who were close behind us. No one wanted to listen, but I knew that Willis knew because I had taught Willis.

Willis hit me again. This time with all of his fingers extended and joined, like a drill sergeant, he pointed at my face. The resolve in his voice told me everything that I needed to know.

"That's the fucking enemy Sir. I don't care what they're saying. That's the fucking enemy and I'm telling you!"

This was IT.

I was an Officer, I took my role seriously. With just ten seconds to decide, one wrong move could mean having our headquarters section would be wiped out.

Willis yelled again.

My training had prepared me to act swiftly and decisively, yet I found myself in an unexpected, high-stress situation with everything on the line. I learned in that moment that I had to take Courage and trust myself in deciding upon a perilous course of action with limited tools and time.

"It's behind us now!"

I had to choose between listening to all of the other Sergeants who were screaming into the radio for me not to shoot, or Private First Class Willis; a young soldier who had been pushed to the brink. He was a guy that nobody wanted...

...Except me.

I shot them. The first bullet bounced off of them! I shot them again and the next bullet bounced off!

He was getting closer to the medics. I was in a panic. The High Explosive Anti Tank (HEAT) rounds my main gun was firing was not stopping their track. Those guys were supposed to be fucking dead. But no luck, my ammo wasn't strong enough.

"Sabot! Sabot!"

Willis yelled at me to change the type of bullets.

He tried to change the bullets and the gun feed-tube jammed! Willis couldn't unjam it. He never had to do this. He didn't even know how to shoot the cannon. Willis wasn't a gunner yet.

Willis and I were the only 2 soldiers in the turret at this point. Dial, the driver, and another soldier were below in the hull. The turret operated like a cylinder with an opening. If you turned the cylinder which we had done, you could not get out. You had to either climb out of the hatch, which we could not do with an advancing enemy tank bearing down on us, or try and climb out of the back of the tank, which we couldn't do.

We had to fight.

Bullets ripped through the air around us. A few ricocheted off of the metal on the Bradley. . I dropped to my knees in the turret, closed my eyes and remembered the First Platoon's Platoon Sergeant Jeffrey Byrant Mitchell's lesson to me on how to unjam a dual fed 25MM Chain gun. Staff Sergeant Mitchell taught me what the Scout Platoon Leader School did not: how to fix the weapon with limited access to tools and no time to make a mistake. Jeffrey was always prepared and enjoyed the respect of all soldiers. While I was in my turret fixing my jammed feed tube, Staff Sergeant Mitchell was also receiving fire from the enemy himself.

In seconds, I unjammed the feed tube with some tools and I could see in my mind Staff Sergeant Mitchell smiling and saying "See Lieutenant, I told you, you could do it" His southern accent and gentlemanly manner were unique.

I changed the type of ammo. I reacquired the target and shot once,

"LONG!" Willis shouted, informing me that the round went over the top of the BMP,

I shot again.

"SHORT!" Willis yelled, and that was magic.

The next round went through the engine and the front sprockets locked up. People flew off of the top of the turret of the Iraqi tank.

Then they got up and started to organize themselves. They had no intentions of peacefully surrendering to us. I opened fire with the machine gun.

But they didn't stop!

I tried to fire the co-ax machine gun again and it jammed.

I went back to the main gun, indexed HEAT ammo and I shot the guy in the front twice in the hips. The first round went off and I didn't realize that it hit him on his right side and went through him. His body was still standing there.

He was still standing. I squeezed another round off and shot him again and both glutes were blown out.

The few survivors raised their hands to surrender. I knew that I had killed enemy soldiers at this point until I heard the words that I will never forget.

Lt. Blake Wallace Radioed me "Cease fire. Cease fire! You just shot Sergeant Thomas!"

All of the air escaped my lungs. I could instantly feel my face go flush as my blood stopped flowing through my body. Horror and anguish fought to be the prevailing emotion in my mind.

I had shot up a tank of our soldiers.

I believed this was possible because I knew that everyone else had been on the briefing with the Captain. They were privy to information that I simply didn't have. Was I too tired? Did I make the unthinkable mistake at a crucial moment? Questions instantly flooded my thoughts.

The depression was instantaneous. I would have to grapple with the thought that I had not only killed two men, I had killed my own men. I shook as if I was freezing as snot drained from my nose. I sobbed like a father, who had lost his only child.

"You killed an American soldier!"

"You fucking idiot!"

"We told you not to shoot!"

"We told you to stand down!"

Everyone yelled at me at once. I had no answers. I had no response. I looked at Willis, but surprisingly he didn't look like me. He was stoic. Almost resolute.

The Captain came up and screamed.

"How could you have done this? How could you have embarrassed the 7th Cavalry Regiment!"

I was already at the end of my line. There wasn't any lower for me to go.

"How could you let the enemy get behind the fucking screen line!"

I said, "What the fuck are you talking about sir?"

He looked me in the eye and then slapped me with an open palm across the face. I barely even felt it because I was still numb from the previous few minutes. But the truth is he wasn't happy that I had saved our platoon and the lives of our men. He was more upset that someone had gotten through our screen line. In his mind, his reputation was bigger than all our lives at that moment.

He had just fucked up his whole career and he knew it the moment he slapped me. His career was over. He looked around at everyone who was standing there and started to walk away.

"Go clean up your fucking mess Lieutenant."

He stormed off to face an uncertain professional future.

Blake Wallace, who was the asshole Lieutenant that told me that I shot Sergeant Thomas. But I hadn't. They had all been wrong. All the shit that they had called me. All of it was absolute Bullshit. They were all wrong and …

…Willis was right.

I drove over to the scene to clean up. My head was reeling from all of this. I had saved my platoon and most certainly the medics and Willis had saved me. Our lives would be linked forever.

A call came out on the radio. It was Lt. Reeves. Reeves was on the left flank and was the most junior officer we had.

Second Lieutenant Reeves keyed his microphone and asked no one in particular: "They're shooting at us. What do we do?"

In the radio background, and in the air around us, we could all hear machine gun fire. It sounded like things were getting hairy. What do we do? That question sounded foreign and illogical to me. This was the way some soldiers were though.

We were in war, and he felt like even though he and his soldiers might be in peril, he needed to ask what he should do.

"Shoot back Lieutenant."

At least my 51-year-old Platoon Sergeant, Sgt V.A. Caminos knew and told him so.

When I arrived at the enemy BMP the man was dying. It's a painful sight to watch someone die. Blood sputtered out of his mouth. He looked at me, as if to say "You're the one who shot me?" I didn't know what to do. Then he died.

We fixed the wounded and we got him secured and there was a prisoner, who I had to drive back to where the medics were.

As we drove back, I could feel a slight sense of pride welling up inside of me. Then I looked and I saw Sergeant Sampson from the headquarters detachment, in his white underwear, no trousers and his tank tee shirt and a pair of untied boots, running towards us. He had run 300 yards through the thick sand towards a combat area with nothing, not even his M-16.

He ran in front of the My Bradley and stopped. I stopped. He looked me in the eyes and then saluted me and stood at attention, as we passed.

I had just written an Article 15 against him for disobeying a direct order weeks before, for a terrible decision that he had made, but he respected me enough to show that sort of gratitude. I never forgot that moment.

I helped turn in the wounded, feeling like a complete wreck—a tangled mess of emotions. The Lieutenants tried to find a way to smooth everything over, but that was a lot to recover from. The Captain switched Lieutenant Reeves and I as Platoon Leaders. I was now Platoon Leader of the Second Platoon with Staff Sergeant Mitchell as my new Platoon Sergeant.

The next day General Tilelli and General Tommy Franks flew in, in a helicopter. I was standing there at the BMP that I shot. I was still a mess. The driver and the other guy had both been killed.

General Franks looked around.

"Anybody kill anybody here?"

"Nobody here Sir."

The Captain looked earnestly at the Generals. In that instant, he erased what I had done. That meant the difference between a medal for me and my soldiers for that action and no medal for me or my troopers. That Captain would've had to explain what happened there if he said "yes"

And in that moment, he disgraced me and my troopers to protect himself.

For the record, Sergeant Thomas and his men were fine. The Lieutenant had never checked to see where Sergeant Thomas was. Thomas's Bradley threw its track off his vehicle when he was maneuvering. So he wasn't where he was supposed to be.

The enemy had driven exactly between where Sgt. Thomas was supposed to be.

No one ever apologized. General Fred Franks noted my combat-action in his book: Into The Storm A Study In Command on page 438.

Everyone knew it. Even soldiers who weren't there like my good friend Michael Reiber. He had heard it play-by-play on the radio. Michael has been there for me for nearly 4 decades and he knew it was a shit show when it happened.

I never received any recognition for what I had done, but more importantly, neither did Private First Class Willis. He had in fact saved us all. And he truly deserved to get his due.

I wrote up a Bronze Star commendation for him and even sent it to Congresswoman Debbie Lasko, but nothing ever came of it.

I wrote two Bronze Star applications for Staff Sergeant Mitchell, one for meritorious service, based on his training an officer, outside of his own platoon which resulted in the protection of our soldiers. The second for his actions exposing himself to incoming fire by using his ANPVS-7 night vision attempting to locate the incoming enemy fire and direct his troops to counter attack. Captain Blockhus never advanced either award.

I regretted my actions with Michael Dial. The incident played over in my head every night. I put him up for a Bronze Star as well, but it was denied.

When my officer's evaluation report was submitted, the Captain wrote that I personally engaged the enemy and thwarted an attack against the Squadron. He wrote it there because it was true, and Colonel Skip Sharp, the 1st Squadron 7th Cavalry Commander, made him write it there.

I survived that night, but a part of me died and I would never get it back.

It occurred to me the next morning: unlike movies: there's no music before the battle begins.

Chapter 19. The Aftermath

The Gulf War ended the next day on February 28, 1991.

The news was bittersweet. I was happy that I was alive and would get to go home and spend time with Catherine, but there was this sense of incompletion. We had spent our military lives preparing for war and just like that the war was over. No one else in our platoon had ever truly understood what it was like to engage in a life-or-death battle, so they looked to me for details.

Everyone knew about my firefight. It gave me some level of notoriety and at that time I didn't mind talking about it. I was doing my best to maintain my image and composure as an officer, but in truth, I was hurting inside.

"What was it like Roux? How did you decide with all that noise and still fire on them?"

It was mostly young soldiers who would ask.

"I had to trust myself" or "I'm damn sure glad Willis studied those enemy tank flash cards. He made it happen just as much as I did." And I always ended it with:

"Always remember that the smallest details in training can be the difference between life and death. This time, it was the training I gave my soldiers that saved us. Train yourself and your troops thoroughly, and kill the enemy quickly, because if you hesitate, he won't."

I had been ostracized from leadership in the 1st Squadron 7th Calvary after that firefight. Though they never said this, and I didn't suffer any ill effects from this move, my platoon had essentially been taken away from me.

Staff Sergeant Mitchell changed my life—not just through his training but through the example he set, even in those brief, pivotal moments we shared.

Jeffrey Mitchell was an emotional giant. In those moments of monumental stress, his readiness freed him to support those around him, even as he faced the same dangers.

His counsel brought me back from a terribly dark place and allowed me to eventually return home and raise a loving family, a debt I cannot measure, nor ever repay.

Some officers spread rumors to justify the move, covering their own backs and inflating their own importance.

I had no idea.

We stayed in the Gulf region until April and then made our way back home. I ended up on the cover of the August 26, 1991 edition of Air Force Times, though of course there was no mention of my firefight. Still, it was good to see the cover and I shared it with my family and friends. My brother, Joe, was especially intrigued by my war stories. He ate up every bit of it, but I found that something was happening to me. I started to retreat from people and felt myself slipping into what I could only describe as some kind of depression.

Soldiers returning from war automatically received 30 days of leave and then an additional 90 days off "to "deprogram." This essentially meant you were left to your own devices and coping methods. My nightmares began immediately. Catherine did her best to comfort me, but I felt like I wasn't in control. I couldn't sleep.

I played the scene over and over in my head. "What could I have done differently?" Should I have not fallen asleep?" "What if Willis had been wrong?"

There was no true reasoning to my questioning and though I had been correct in my thinking, I still had this sense that I had done something wrong. I had seen Sergeant Thomas regularly, but I still could not escape the "What if?" This question haunted me for many years and hurt many of my personal relationships.

My depression worsened. I tried to get out of it, but it simply would not happen. For a proud officer and soldier, for the first time in my life, I felt helpless and even ashamed. I talked to doctors, but they couldn't figure out what was wrong with me. Post Traumatic Stress Disorder recognition and diagnosis were still in their infancy. I had PTSD and no one, most of all me, knew it.

It would take over 25 years for me to finally be diagnosed and be afforded the resources and treatments that I so desperately needed.

I got my last promotion to 1st Lieutenant on January 25, 1992, and received the Army Commendation Medal 2 weeks later. I received a medical Honorable Discharge on February 18, 1992.

My military career had ended. In war, I had faced death, wrestled with failure, and emerged with scars both visible and unseen. But it was the Honduran jungle—an untamed frontier of greed and peril—that now called me to return again. It demanded a different kind of courage, one that required not only my training but my soul. The weight of this responsibility pressed against me, heavier than any battlefield burden. However, I was unaware of The Mahogany Mafia's deeply entrenched power—its ruthless grip on the jungle, and its willingness to destroy anyone who dared to challenge its reign.

This was the new battlefield. And my story continues in:

Book 2: **The Mahogany Mafia - Murder and Conspiracy in the Jungle**

A Glimpse Ahead

Alacrity And Dispatch: The Chronicles Of A Citizen-Soldier's Selfless Service

Is a groundbreaking five-book non-fiction series that unearths the extraordinary journey of a man whose relentless pursuit of justice and innovation reshaped industries and challenged corruption. Documented with meticulous precision, the series ties together seemingly disparate events—from counting timber in the jungle to exposing systemic corruption in America's power grid—into a cohesive story of courage, ingenuity, and resilience. Every chapter is rooted in firsthand accounts and the author's own technological breakthroughs, illustrating how a singular focus on accountability can ignite transformative change. With no names hidden and no detail spared, the books reveal not only the author's remarkable achievements but also the deeper, interconnected forces shaping our modern world.

Coming Next: The Mahogany Mafia

In the Honduran jungle, the author confronts personal loss and global corruption after the murder of his father and the imprisonment of his staff. This volume uncovers the dark side of the timber trade, where criminal networks exploit natural resources and human lives, setting the stage for the author's life-changing innovations.

Preview: The Mahogany Mafia Chapter 1

Chapter 1 Return Of Gringo Joe

The jungle felt familiar but unforgiving. As I stood beneath the sprawling canopy, I couldn't help but reflect on the journey that had brought me here. War had shaped me, honed my instincts, and taught me to survive under the most harrowing conditions. But nothing in my military training had prepared me for what awaited in the Honduran jungle—the web of corruption, greed, and violence orchestrated by The Mahogany Mafia.

I had thought the scars of war were the heaviest burden I'd ever carry, but I was wrong. Here, survival wasn't just about endurance; it was about navigating a shadowy world where every ally could be a traitor, and every decision could cost lives.

Later that morning, I stood on the balcony of my modest office, staring out over the dense, vibrant jungle that stretched as far as the eye could see. The sun was beginning its descent, casting a golden hue over the landscape. But I had no time to admire the beauty around me. My mind was consumed by the urgent crisis unfolding in my timber business.

A sharp knock on the door broke my concentration. I turned to see my trusted general manager, Jimmy Malone, entering the room, twisting his moustache with a worried expression.

"Adam, we just received word from Olanchito, another shipment has been hijacked. The locals are getting really angry, and our clients are threatening to pull out if we can't secure the deliveries."

I ran a hand through my hair, feeling the weight of the situation pressing down on me. I knew that the stakes were high. This was not just about business; it was about survival in a land where corruption and danger lurked around every corner.

"We need to figure this out Jimmy. Double the security detail and try to negotiate with the local leaders. We can't afford any more losses and neither can they. We are on the same team with them, remind them of that and tell them I am coming to meet with Sr. Maquez, in Olanchito, in person, on Monday" I said, my voice steady but tinged with frustration.

As Jimmy left to find a landline to make the call and execute the plan, my thoughts drifted back to my father, Joe Rousselle, or "Gringo Joe" as he was known in Honduras. Dad had always been a larger-than-life figure, full of charm and audacity. His entrepreneurial spirit had led him to Honduras years ago, where he carved out a niche for himself in the most unexpected of ways.

I remembered the first time I reconnected with my father after joining the Army. Dad was living in a small coastal town, surrounded by people who admired his charisma and business acumen. Joe's life was a stark contrast to the disciplined, structured world I had grown accustomed to in the military.

He lived above the Pharmacia Nueva with Argentina de Batres. Their house was modest but exuded an air of comfort and affluence. I found myself drawn to the stories my father told about his life in Honduras. Gringo Joe had made a name for himself by importing and selling personal hygiene products, a venture that seemed trivial but was surprisingly lucrative in the local market.

"Adam, you've got to understand something," he said one evening over dinner. "Opportunities are everywhere if you know where to look. You've just got to attack them at the right moment."

His words were nearly the same as we used in combat, perhaps he intended so. His words served as a guiding principle in my own entrepreneurial journey. Despite the complexities of our relationship, I couldn't deny the impact my father had on shaping my worldview.

I snapped back to the present as my phone rang loudly on the desk, few folks had this warehouse number, and it was likely from overseas. My heart skipped a beat as I answered.

"Mr. Rousselle, This is the Palmerola Airforce Base operator, hold for a patch from Joseph Rousselle" Adam, it's Dad." Gringo Joe's voice came through the line, as confident and commanding as ever. "I'm back in Honduras. We need to talk." It had the unmistakable tone that someone else was listening.

The timing couldn't have been worse. My business was teetering on the edge, and the return of my father, with all his schemes and ambitions, could either be a lifeline or a disaster.

"Dad, I'm dealing with a lot right now. Jimmy's up here. Can this wait?" I tried to keep my tone neutral.

"No, it can't. I have a idea that could turn things around. Meet me in La Ceiba tomorrow evening." The line went dead.

Dad had a way of pulling me into his whirlwind of plans and ideas, often leading to unpredictable outcomes.

"Fine. I'll be there, I said as the disconnect tone was in my ear." I was already bracing myself for whatever my father had in store.

As I hung up the phone, I couldn't shake the feeling that my father would bring both new opportunities and unforeseen challenges. Gring Joe's presence always stirred the pot, and in the volatile environment of Honduras, I knew firsthand that always meant trouble, for someone.

I landed in San Pedro Sula and drove to la Ceiba. I stood on the balcony above the Pharmacia Nueva once more, looking out past the church and at Pico Bonito and the jungle I used to sleep in. The sun had set, and darkness was creeping in. I knew that the days ahead would test my resilience and resolve in ways I hadn't yet imagined.

But if there was one thing I had learned from the 7th Special Forces, it was to face challenges quickly and with violence-of-action. With a deep breath, I steeled myself for the next chapter of my life, where the lines between business and survival would blur once again.